Automation, Collaboration, & E-Services

Volume 7

Series Editor

Shimon Y. Nof, PRISM Center, Grissom, Purdue University,
West Lafayette Indiana, IN, USA

The Automation, Collaboration, & E-Services series (ACES) publishes new developments and advances in the fields of Automation, collaboration and e-services; rapidly and informally but with a high quality. It captures the scientific and engineering theories and techniques addressing challenges of the megatrends of automation, and collaboration. These trends, defining the scope of the ACES Series, are evident with wireless communication, Internetworking, multi-agent systems, sensor networks, cyber-physical collaborative systems, interactive-collaborative devices, and social robotics – all enabled by collaborative e-Services. Within the scope of the series are monographs, lecture notes, selected contributions from specialized conferences and workshops.

More information about this series at http://www.springer.com/series/8393

Itshak Tkach · Yael Edan

Distributed Heterogeneous Multi Sensor Task Allocation Systems

 Springer

Itshak Tkach
Rishon LeZion, Israel

Yael Edan
Department of Industrial Engineering
and Management
Ben-Gurion University of the Negev
Be'er Sheva, Israel

ISSN 2193-472X   ISSN 2193-4738   (electronic)
Automation, Collaboration, & E-Services
ISBN 978-3-030-34737-6   ISBN 978-3-030-34735-2   (eBook)
https://doi.org/10.1007/978-3-030-34735-2

This Springer imprint is published by the registered company Springer Nature Switzerland AG
The registered company address is: Gewerbestrasse 11, 6330 Cham, Switzerland

I wish to dedicate this book to my parents,
Alexander and Irina Tkach.
Itshak

Foreword

Integration and collaboration of sensors, communications, and tasks are critical yet not well investigated in today's widespread Internet of things (IoT) applications. This book provides a clear pathway for the readers to understand the sophisticated systems which include two functions: (1) allocating sensors to tasks and (2) task administration to maintain the functionality of the sensor networks.

Although building an abundant number of sensors in a system and trusting the sensors can guarantee the health of the system become a norm today, not many studies have been done to realize the robustness of the system and the association of the sensor network with the dynamic task requirements of the uncertain system. The pioneering study was initiated in Production, Robotics, and Integration Software for Manufacturing and Management (PRISM) Center at Purdue University in the 1990s. At that time, there was neither IoT nor cloud technology. After so many years, I am pleased that Itshak Tkach and Yael Edan wrote this book to demonstrate a full set of theories and applications. The theories and applications are especially meaningful and important for today's IoT world.

The authors successfully constructed and proved the necessary theories of the multi-sensor network. With the demonstrations in this book, the readers can find solid reasons for deploying and managing the sensors with a certain specification in their domain applications. When I congratulate the extraordinary achievements the authors had done, in the meantime I thank them for giving us a solid step stone to further investigate the world of multi-sensors.

Chin-Yin Huang
Tunghai University, Taiwan

PRISM Center, Purdue University, USA

International Foundation for Production Research
Taichung, Taiwan

Preface

Heterogeneous sensor system research can gain from exploring approaches commonly found in swarm intelligence and task administration protocols. This book includes the description of a distributed algorithm for efficient and scalable heterogeneous multi-sensor task allocation and task administration protocols (TAPs) to handle problems in the process layer of the system.

This book presents a double-layer heterogeneous multi-sensor system for target allocation (DL-TDS), which includes a process layer that allocates sensors to tasks and a monitoring layer that is dedicated to handle special cases of problems within the sensory system of very long attendance times, priority conflicts, and failure. The two-layer system deals with heterogeneous sensors with different performances that are distributed a priori in the area of interest (=the area in which tasks occur). It was used to allocate multiple tasks, with unknown a priori priorities that arrive at unknown locations at unknown times. The process layer—PL—uses a bio-inspired swarm intelligence heterogeneous distributed bees algorithm (HDBA) that was developed for heterogeneous sensor allocation yielding efficient and scalable performance. It uses a dynamic temporal and spatial allocation of sensors to tasks with different priorities appearing at different times and locations and decides which sensor to allocate to which tasks, when, and where. For the monitoring layer—ML, four task administration protocols (TAPs) were implemented to overcome uncertainties in task arrival and sensory performance and disturbances (i.e., high time-consuming tasks, conflicts in task priorities, and sensor failure, all defined as overloading, deception, and tampering of sensors, respectively) in the multi-sensor system in the process layer. The developed protocols also ensure optimal sensor availability related to their monetary cost in the system. "TRAP" manages task priorities to allow better allocation of sensors to the most important tasks and to detect tasks when they occur. "ASAP" ensures that tasks will be treated as soon as possible and will not be unnecessarily delayed. "SRAP" allocates the best match of sensors to treat tasks, and "STOP" was designed to ensure optimal sensor availability in the system by time-out policy. Employing TAPs with HDBA allows dynamic, real-time allocation of distributed sensors to tasks when they occur. The system reacts to dynamic task occurance and applies protocols along execution and according to internal sensors' performance using an objective system function developed to predict system performance given the sensor, environmental, and task parameters.

In the extended example chapters, HDBA was evaluated in simulation in comparison with four other state-of-the-art algorithms (DBA, bees system, market-based, and greedy). Three different deployments were analyzed, grid deployment, uniformly distributed random deployment, and normally distributed random deployment. The algorithm efficiently assigns a heterogeneous swarm of sensors to upcoming tasks by providing scalability, in terms of the number of tasks and sensors. The HDBA resulted in significantly better system performance in terms of both allocation times and the number of unallocated tasks in comparison with other algorithms.

Additional evaluations of HDBA were conducted on a benchmark traveling salesman problem (TSP) for two different cities and five algorithms, and on a law enforcement problem (LEP) in comparison with the FMC_TA^{H+} algorithm and simulated annealing algorithm. Results indicated HDBA's fitness to solve the TSP and ability to allocate heterogeneous police officers in LEP.

The performance of the dual-layer system was simulated for a wide range of different scenarios, and the results indicated statistically significant improvement in performance of up to 72% of the number of allocated tasks compared to a solely operating allocation algorithm.

System reliability was evaluated using Monte Carlo simulations for the heterogeneous sensor network. Simulation analyses indicate that overall systems' availability was improved with statistical significance of 95%, ensuring fault-tolerant system operation. Simulation results of TAP operation indicated a statistically significant increased number of processed tasks (by 13.1%, $p < 0.05$%) when reliability analysis recommendations were applied.

Rishon LeZion, Israel   Itshak Tkach
Be'er Sheva, Israel   Yael Edan

About This Book

Today's real-world problems and applications in target detection require an efficient, comprehensive, and fault-tolerant multi-sensor allocation system. This book provides theory and applications of novel methods developed for multi-sensor systems. Advances in multi-agent systems and AI along with collaborative control theory and tools are used within this book. A dual-layer task allocation system that uses a new swarm intelligence algorithm for heterogeneous sensors and protocols to overcome problems of overloading, deception, and tampering is described and explained. It presents the formulation and development of an allocation framework for a heterogeneous multi-sensor system for different real-world problems that require sensors with different performances to allocate multiple tasks, with unknown a priori priorities that arrive at unknown locations at unknown time. It explains how to decide which sensor to allocate to which tasks, when, and where. Reliability and availability issues of task allocation systems are also explained, and methods for their optimization are given.

The following features are enabled by the decentralized architecture described within this book:

1. Robustness to sensor failure (fault tolerance)—an ability to continue operation despite partial failure in system components,
2. Quick response to dynamic conditions,
3. Efficient allocation of system members (sensors)—the efficiency is defined by performing allocation in minimal possible time,
4. Allocation of sensors to tasks in real time when information is not known a priori—allocation is an amount or portion of a resource assigned to a particular task,
5. Ability to deal with limited range coverage of sensors,
6. Ability to allocate a limited number of sensors to treat dynamic tasks,
7. Ability to dynamically handle new tasks,
8. Ability to dynamically reallocate sensors to tasks,

9. Ability to accommodate addition/subtraction of sensors during operation (scalability),
10. Ability to accommodate heterogeneous sensors, and
11. Ability to allocate large numbers of sensors.

These features are explained, measured, and evaluated by extensive simulations, and the results of these simulations are presented in this book.

This book will appeal to academics, researchers, and graduate students as well as engineers and professionals, and is relevant to various applications such as multi-agent systems, task allocation, optimization, target allocation, team formation, sensor network design, facility monitoring in industry, cyber security, fire monitoring, surveillance, and homeland security among others.

Contents

Chapter 1
Introduction

This chapter presents the fundamentals of distributed heterogeneous multi-sensor task allocation systems with several examples and illustrations. The purpose is to present the basic definitions of these complex systems in the context of multi-agent, scalable, and reliable systems, highlight its impact on the competitive performance of such systems, and outline the structure and objectives of this book.

1.1 Sensory Task Attendance and Allocation

Real-time allocation of dynamic tasks with unknown a priori locations, priorities, and time of occurrence, are a common problem in different applications [7, 24, 36]. Multi-sensor systems can provide a robust solution for attending different tasks, overcoming the limitations inherent to a single sensor [18, 22] and yielding improved performance [21, 23, 36].

Task attendance by sensors is defined by Robin and Lacroix [23] as finding a task in a given environment. Tasks may be attended by single or multiple sensors, which can be either mobile or fixed static sensors [23]. The assumption is that correct task allocation requires a certain amount of attendance time and hence is not instantaneous. Multi-sensor systems include distributed and centralized systems [35] with homogenous or heterogeneous sensors with different features (e.g., response time, resolution, field of view).

Recent progress in sensor technologies, especially for small scale sensors, provides the ability to form swarms with advanced capabilities [4]. Research in multi-sensor systems [14] deals with sensor selection, allocation [12], disturbances [26], and scheduling [12, 31]. The selection problem refers to the decision of which sensors will be used to attend the tasks. The allocation problem deals with the assignment of available sensors to tasks to maximize the total utility of the system [12]. Scheduling deals with the order in which the task requests will be executed, aiming to improve the overall mean response time to tasks [12]. Multi-sensor task allocation can be

© Springer Nature Switzerland AG 2020
I. Tkach and Y. Edan, *Distributed Heterogeneous Multi Sensor Task Allocation Systems*,
Automation, Collaboration, & E-Services 7,
https://doi.org/10.1007/978-3-030-34735-2_1

defined as a variation of a multi-robot task allocation problem (MRTA) as defined by Gerkey and Matarić [9].

This book focuses on developing robust scalable methods for sensor allocation with the ability to accommodate addition/subtraction of sensors during operation and handle problems existing in allocation methods for task attendance with distributed heterogeneous sensors. The tasks (which can also be defined as targets) arrive at unknown times and locations and have different priorities, which are related to the task's importance. Tasks with higher importance must be attended faster than other tasks and obtain a greater benefit when they are detected. The aim is to attend all tasks as soon as possible. The system considers each task as a task that requires attendance by a sensor for predefined time durations. The attendance time depends on the task type and the sensors allocated to it. A task is considered as attended when the corresponding time has been performed by the sensor at the task. Each sensor has different performance features (e.g., allocation distance, resolution), which are defined a priori according to the sensor type. The aim is to ensure continuous allocation despite sensory malfunctions. This system is adaptive to dynamic changes in task occurrences (i.e., able to allocate and reallocate sensors to new tasks, which are dynamic in place and time), and is scalable (i.e., easy to add or remove sensors) to cope with the increasing number of tasks.

1.2 Basic Definitions

Oxford dictionaries define **sensor** as "A device which detects or measures a physical property and records, indicates, or otherwise responds to it". In other words, a sensor is a device that detects and responds to some type of input from the environment. The specific input could be light, heat, motion, moisture, pressure, or any one of a great number of other environmental phenomena. The output is generally a signal that is converted to a human-readable display at the sensor location or transmitted electronically over a network for reading or further processing.

According to the Merriam-Webster dictionary, a **task** is "a usually assigned piece of work often to be finished within a certain time". A single task can represent a security event, a crime incident, fire, a leak in a pipeline and any other assignment that must be detected, monitored or carried out within a defined time slot.

Scalability is a characteristic of a system that allows it to cope with the increased or decreased quantity of agents or tasks. This ability is also defined as scaling-up or scaling-down.

1.3 The Need for Efficient Multi-sensor Task Allocation

In applications that require constant detecting of targets (tasks), distributed multi-sensor systems can play an important role due to their capability to cover the entire area and ensure a robust response to dynamic situations [2]. Most task allocation multi-sensor systems rely on a very large number of sensors, usually swarms, to cover the entire area and to be able to allocate tasks [1, 33, 36], which makes the problem of the efficient allocation of sensors to tasks NP-hard [32, 36]. Each sensor has different capabilities (such as allocation distance, resolution). When there are several tasks that require the same sensors for allocation, a decision must be made regarding which sensors should be allocated to which task. The allocation depends on the sensors' availability and performance, and the priorities of individual tasks. This problem becomes more complicated when tasks arrive at unknown locations and unknown times, and sensors are heterogeneous. The problem of sensor disturbances deals with external and internal factors that can affect system performance (i.e., reliability of the hardware, weather conditions, jamming; [29]). This system requires solutions that can offer a rising level of robustness and efficiency [39].

There are different types of algorithms for sensor allocation including centralized, decentralized, and hybrid allocation algorithms ([35], Chap. 2). However, in situations that include problems within the sensory system such as deception, overloading, and tampering (described in detail in Chap. 7), algorithms for sensor allocation cannot cope solely with these problems to effectively allocate tasks.

In this book, a dual-layer task allocation system (DL-TDS) is presented to overcome these problems by developing an efficient allocation algorithm and coordination protocols. It includes (1) a process layer—PL (multiple sensors and allocation algorithm HDBA for allocating them to efficiently complete tasks); (2) a monitoring layer—ML (task administration protocols for handling problems in the sensory system in the process layer). The DL-TDS is relevant for various applications such as facility monitoring in industry (e.g., detecting and preventing errors, security threats, and thefts, [13, 30, 41, 42]), cyber security (e.g., protecting systems from cyberattacks and providing near real-time response, [6, 8, 10]), fire monitoring (i.e., 24 h forest fire protection to react fast enough to suppress fire occurrence or to minimize damage made by forest fires, [5, 16, 17, 38, 40]), surveillance [20, 27], transportation security [11, 37], military (i.e., detecting moving enemy targets, monitoring soldier camps, [3]), and homeland security (airports, railroads, and highways as well as water, power, and energy sources security monitoring, [15, 19, 25, 28, 34]) among others.

1.4 Main Aims of This Book

The objective of this book is to present a framework for allocating heterogeneous sensors to multiple tasks, that arrive at unknown locations at unknown times with unknown a priori priorities. The book will illustrate how to:

1. Compute the expected value of system performance given the sensor, environmental, and task parameters.

2. Decide in real time which sensor addresses which task without knowing a priori when, where, and which task will enter or leave the environment.
3. Handle risks and problems within the sensory system such as sensor availability, conflicts in tasks priorities, and high-time consuming tasks.
4. Ensure fault tolerant sensor operation that is robust enough to sensor failure.

Figure 1.1 illustrates the workflow process of multi-sensor task allocation that will be described and explained throughout this book.

1.5   Book Summary

This book presents a heterogeneous multi-sensor system for task allocation applications. It provides a framework to enable efficient allocation of heterogeneous sensors to multiple tasks arriving at unknown different times and locations and to handle problems within the sensory system. An objective function of sensor performances based on the tasks' priorities and the distances of the sensors from tasks to quantify system performance is presented. A Heterogeneous Distributed Bees Algorithm (HDBA) that uses the principles of swarm intelligence is applied for efficient heterogeneous sensor allocation. It uses a dynamic temporal and spatial allocation of sensors to tasks with different priorities appearing at different times and locations and decides which sensor to allocate to which tasks when and where. Four task administration protocols are used to handle problems arising in the multi-sensor system. "TRAP" manages task priorities to allow better allocation of sensors to the most important tasks and to detect tasks when they occur. "ASAP" ensures that tasks will be treated as soon as possible and will not be unnecessarily delayed. "SRAP" allocates the best match of sensors to treat tasks and "STOP" was designed to ensure optimal sensors availability in the system by time-out policy. These protocols are part of DL-TDS which is implemented for several case studies in order to simulate and validate its performance. It is based on a dual-layer architecture, including a process layer and a monitoring layer. The process layer consists of multiple sensors and is responsible for allocating them to complete tasks. The monitoring layer is used to monitor problems in the process layer and applies task administration protocols for handling them. This architecture uses a decentralized approach for task allocation.

This book is divided into 11 chapters. Chapter 2 presents the different approaches to multi-agent task allocation. Chapter 3 describes the multi-agent task allocation systems and their framework. Chapter 4 illustrates the evaluation methodology applied in this book. Chapter 5 describes the single-layer multi-sensor task allocation system and algorithms for multi-sensor task allocation. Chapter 6 deals with extended examples of the single-layer multi-sensor task allocation system. The examples include implementation of the single-layer multi-sensor task allocation system to the security of supply networks, the traveling salesman problem and the law enforcement problem. Chapter 7 describes the dual-layer multi-sensor task allocation system. It

Fig. 1.1 Sensors allocation
workflow scheme. Sensors
are represented as agents,
either homogeneous or
heterogeneous. Agents are
allocated to tasks by
allocation algorithms. If
there are disturbances—task
administration protocols are
assigned

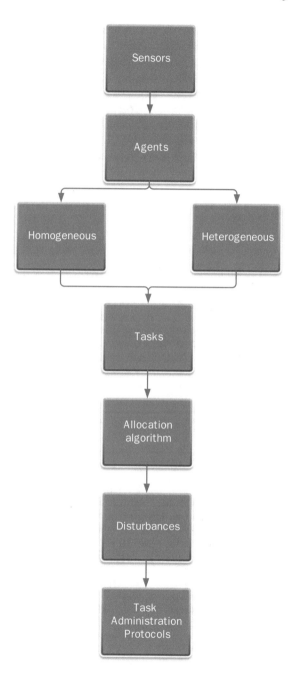

presents the framework design and the formulation of task administration protocols. Chapter 8 presents extended examples of the dual-layer system and evaluations through numerical analyses. Chapter 9 presents the development of a fault tolerant multi-sensor system with high availability. It defines the system's availability, presents the reliability design, and evaluations through Monte Carlo simulations. Chapter 10 presents an analytical analysis of a simplified scenario of two sensors and two tasks for single-layer and double layer systems. The book concludes in Chap. 11, with a summary, final remarks, and discussion of future research directions to extend the concepts and framework presented in this book.

References

1. Akyildiz IF, Su W, Sankarasubramaniam Y, Cayirci E (2002) Wireless sensor networks: a survey. Comput Netw 38(4):393–422
2. Anastasi G, Conti M, Di Francesco M, Passarella A (2009) Energy conservation in wireless sensor networks: a survey. Ad Hoc Netw 7(3):537–568
3. Ball MG, Qela B, Wesolkowski S (2016) A review of the use of computational intelligence in the design of military surveillance networks. In: Recent advances in computational intelligence in defense and security. Springer International Publishing, pp 663–693
4. Bayındır L (2016) A review of swarm robotics tasks. Neurocomputing 172:292–321
5. Bernardo L, Oliveira R, Tiago R, Pinto P (2007) A fire monitoring application for scattered wireless sensor networks. In: Proceedings of the international conference on wireless information networks and systems, Barcelona, Spain, vol 2831
6. Böhme R, Schwartz G (2010) Modeling cyber-insurance: towards a unifying framework. WEIS, Harvard, USA
7. Civelek M, Yazici A (2016) Automated moving object classification in wireless multimedia sensor networks. IEEE Sens J 17(4):1116–1131
8. Fan Y, Zhang L, Du Y (2019) A new type building fire protection facility monitoring system. In: International conference on applications and techniques in cyber security and intelligence, June. Springer, Cham, pp 78–87
9. Gerkey BP, Matarić MJ (2004) A formal analysis and taxonomy of task allocation in multi-robot systems. Int J Robot Res 23(9):939–954
10. Haack JN, Fink GA, Maiden WM, McKinnon D, Fulp EW (2009) Mixed-initiative cyber security: putting humans in the right loop. In: The first international workshop on mixed-initiative multiagent systems (MIMS) at AAMAS
11. John A, Yang Z, Riahi R, Wang J (2018) A decision support system for the assessment of seaports' security under fuzzy environment. In: Modeling, computing and data handling methodologies for maritime transportation. Springer, Cham, pp 145–177
12. Kapoor KN, Majumdar S, Nandy B (2015) Techniques for allocation of sensors in shared wireless sensor networks. J Netw 10(01):15–28
13. Lee CKH, Ho GTS, Choy KL, Pang GKH (2014) A RFID-based recursive process mining system for quality assurance in the garment industry. Int J Prod Res 52(14):4216–4238
14. Lekidis A, Stachtiari E, Katsaros P, Bozga M, Georgiadis CK (2018) Model-based design of IoT systems with the BIP component framework. Softw Pract Exp 48(6):1167–1194
15. Lewis TG (2006) Critical infrastructure protection in homeland security: defending a networked nation. Wiley
16. Li Y, Wang Z, Song Y (2006) Wireless sensor network design for wildfire monitoring. In: 2006 6th IEEE world congress on intelligent control and automation, June, vol 1, pp 109–113

17. Liu Y, Liu Y, Xu H, Teo KL (2018) Forest fire monitoring, detection and decision making systems by wireless sensor network. In: 2018 Chinese control and decision conference (CCDC), June. IEEE, pp 5482–5486

18. Mercuri M, Rajabi M, Karsmakers P, Soh PJ, Vanrumste B, Leroux P, Schreurs D (2015) Dual-mode wireless sensor network for real-time contactless in-door health monitoring. In: IEEE MTT-S international microwave symposium (IMS), pp 1–4

19. Ostfeld A, Uber JG, Salomons E, Berry JW, Hart WE, Phillips CA, di Pierro F (2008) The battle of the water sensor networks (BWSN): a design challenge for engineers and algorithms. J Water Resour Plan Manag 134(6):556–568

20. Paramanandham N, Rajendiran K (2018) Multi sensor image fusion for surveillance applications using hybrid image fusion algorithm. Multimed Tools Appl 77(10):12405–12436

21. Pennisi A, Previtali F, Gennari C, Bloisi DD, Iocchi L, Ficarola F, Vitaletti A, Nardi D (2015) Multi-robot surveillance through a distributed sensor network. In: Cooperative robots and sensor networks. Springer International Publishing, pp 77–98

22. Rak MB, Wozniak A, Mayer JRR (2016) The use of low density high accuracy (LDHA) data for correction of high density low accuracy (HDLA) point cloud. Opt Lasers Eng 81:140–150

23. Robin C, Lacroix S (2015) Multi-robot target detection and tracking: taxonomy and survey. Auton Robots 1–32

24. Schneider E, Sklar EI, Parsons S, Özgelen AT (2015) Auction-based task allocation for multi-robot teams in dynamic environments. In: Towards autonomous robotic systems, pp 246–257

25. Shrivastava S, Adepu S, Mathur A (2018) Design and assessment of an orthogonal defense mechanism for a water treatment facility. Robot Auton Syst 101:114–125

26. Tang Y (2016) Coordination of multi-agent systems under switching topologies via disturbance observer-based approach. Int J Syst Sci 1–8

27. Tang Z, Ozguner U (2005) Motion planning for multitarget surveillance with mobile sensor agents. IEEE Robot 21:898–908

28. Turoff M, Chumer M, Hiltz SR, Klashner RM, Alles M, Vasarhelyi M, Kogan A (2004) Assuring homeland security: continuous monitoring, control and assurance of emergency preparedness. J Inf Technol Theory Appl (JITTA) 6(3):3

29. Walters JP, Liang Z, Shi W, Chaudhary V (2007) Wireless sensor network security: a survey. In: Security in distributed, grid, mobile, and pervasive computing, vol 1, pp 367

30. Wang H, Chen S, Xie Y (2010) An RFID-based digital warehouse management system in the tobacco industry: a case study. Int J Prod Res 48(9):2513–2548

31. Wang J, Qin J, Ma Q, Kang Y, Fu X (2018a) Optimal sensor scheduling for two linear dynamical systems under limited resources in sensor networks. Neurocomputing 273:101–110

32. Wang J, Wang Y, Zhang D, Wang F, Xiong H, Chen C, Qiu Z (2018b) Multi-task allocation in mobile crowd sensing with individual task quality assurance. IEEE Trans Mob Comput 17(9):2101–2113

33. Wang P, Yang F, Zhang Y, Zhang L (2018c) Multi-sensor and multi-target task allocation method based on improved firefly algorithm. In: Global intelligence industry conference, vol 10835. International Society for Optics and Photonics

34. Wise CR (2006) Organizing for homeland security after Katrina: is adaptive management what's missing? Public Adm Rev 66(3):302–318

35. Yan WQ (2016) Introduction to intelligent surveillance. Springer

36. Yick J, Mukherjee B, Ghosal D (2008) Wireless sensor network survey. Comput Netw 52(12):2292–2330

37. Yoon SW, Velasquez JD, Partridge BK, Nof SY (2008) Transportation security decision support system for emergency response: a training prototype. Decis Support Syst 46(1):139–148

38. Yu L, Wang N, Meng X (2005) Real-time forest fire detection with wireless sensor networks. In: IEEE proceedings of international conference on wireless communications, networking and mobile computing, vol 2, pp 1214–1217

39. Yuan Q, Guan Y, Hong B, Meng X (2013) Multi-robot task allocation using CNP combines with neural network. Neural Comput Appl 23(7–8):1909–1914

40. Zhang J, Li W, Han N, Kan J (2008) Forest fire detection system based on a ZigBee wireless sensor network. Front For China 3(3):369–374
41. Zhang Y, Qu T, Ho OK, Huang GQ (2011) Agent-based smart gateway for RFID-enabled real-time wireless manufacturing. Int J Prod Res 49(5):1337–1352
42. Zhou F, Lin X, Luo X, Zhao Y, Chen Y, Chen N, Gui W (2018) Visually enhanced situation awareness for complex manufacturing facility monitoring in smart factories. J Vis Lang Comput 44:58–69

Chapter 2
Multi-agent Task Allocation

Multi-agent allocation has become a popular area of research and has advanced significantly in recent years in many applications such as multi-robot task allocation, path planning, control of unmanned aerial vehicles, communication networks, conflict and error prevention, and formation of mobile robots [1, 12, 51, 55, 61]. Multi-agent task allocation problems consist of a set of agents and a set of tasks that the agents must execute [31, 57]. According to Gerkey and Matarić [20] and Robin and Lacroix [44], tasks can be divisible, i.e., each task can be performed by an individual or by a group of agents, and may also require collaboration between agents. The problems of task allocation considered in the literature are mainly multi-agent problems, hence the question of centralized and decentralized systems arises [44]. There are diverse algorithms that are intended to solve task allocation [38, 42, 48, 53, 56]. In general, the multi-agent task allocation approaches can be divided into three categories: centralized, decentralized, and hybrid approaches [60]. The objective function of these approaches is to maximize the overall utility or to minimize the cost of performing the tasks by the agents under a variety of constraints.

2.1 Centralized Multi-agent Task Allocation

Centralized approaches are usually based on a central agent that coordinates the allocation of other agents [36]. These approaches are best suited for applications where teams are small and global information about the tasks is easily available [15]. The primary advantage of centralized systems is that in some problems, the obtained solutions can be optimal or very close to the optimal [36, 44]. However, this is under the assumption that the information from the different agents is accurate enough. But, centralized systems are less robust since they rely on a central element that calculates the optimal allocation and in case of failure of this element, the whole system fails [9, 10]. Other problems related to this approach include limitations with communication coverage (i.e., broadcasting messages long distances from a

© Springer Nature Switzerland AG 2020
I. Tkach and Y. Edan, *Distributed Heterogeneous Multi Sensor Task Allocation Systems*,
Automation, Collaboration, & E-Services 7,
https://doi.org/10.1007/978-3-030-34735-2_2

centralized agent, where large teams of agents cover a large space) and scalability (i.e., requires altering algorithms for addition/subtraction of sensors during operation, [3, 36]).

This book deals with the problem of designing a robust and scalable system with a large number of agents and tasks. These agents are considered to operate in unknown environments without knowing a priori when, where, and which task will enter or leave the environment. This problem cannot be handled by a centralized approach.

2.2 Decentralized Multi-agent Task Allocation

In self-organized decentralized approaches, each agent takes their own decisions without major consideration of other agents [33]. They include methods based on swarm intelligence, such as ant colony or distributed bees, that achieve complex collective behavior from the local interactions of many individuals with simple behavior. In these approaches, sensors use local knowledge and share information with each other [3, 45]. In such systems, sensors collaborate in order to achieve a global goal. Bio-inspired approaches of swarm intelligence, inspired by the emergent behavior of social insects, such as ants, bees, and termites, as well as from other animal societies, such as flocks of birds or schools of fish, have been used to model the behavior of intelligent multi-agent systems [3, 5, 16, 26]. In swarm intelligence algorithms [7, 8, 49], cooperating agents interact by exchanging locally available information, such that the global objective is obtained more efficiently than it would be by agents that perform tasks individually [26]. Swarms can be useful because they can deliver performance that is better than the sum of the parts. These algorithms have been applied to applications of multi-agent exploration and path formation [23], energy optimization in sensor networks [4, 54], multi-site deployment [6], parallel computing optimization [50], task allocation [27], cooperative transport and vehicle routing [34, 63], feature selection [59], intruder allocation [18], resource allocation [41], multi-robot task allocation and tracking applications [25, 47], knapsack problem [11], or cooperative transport and prey retrieval [34]. Disadvantages of swarm intelligent control include conflicts between the members of swarms, redundant activities, and slow global response to a change in the environment [8, 17]; generally they provide only suboptimal solutions [44].

The Distributed Bees Algorithm (DBA) is suitable for implementation in a multi-agent system and is also scalable with low computational overhead consumption [25]. This decentralized topology inherent to bio-inspired multi-agent systems provides them with the ability to be autonomous, scalable, robust, and adaptive to changes in their environment and to real-world constraints [44].

These approaches, although suitable for the multi-agent application, have not been implemented for allocating heterogeneous static agents to dynamic tasks.

2.3 Hybrid Multi-agent Task Allocation

Hybrid approaches use decentralized agents to control and exploit points of centralization in the form of auctions to produce allocations [24, 29]. They include intentional methods such as market-based algorithms and contract net protocols. In such systems, self-interested agents participate in a virtual market economy and allocate tasks by bidding procedures.

2.4 Summary

Various domains and applications of the multi-agent task allocation problem are discussed in the literature. These include agents as robots [21, 22, 39, 40, 58], wireless sensors [46, 62], computers or processors [2, 13, 32, 37], or human agents [28, 42]. For each domain, there is a different objective function, a different set of variables, and a different set of constraints. The task allocation problem can be distributed and dynamic, i.e., new tasks are added to the system over time ([13, 14, 28, 42, 56, 43]). Task allocation problems with a linear objective function and linear constraints can be reduced to an instance of the optimal assignment problem [30]. This is a well-known operations research problem, which can be solved in polynomial time [19]. However, other problem domains, like the problem solved in this book, are nonlinear and complex and have been proven to be NP-hard [21, 35, 52]. These types of problems require incomplete optimization methods like meta heuristics.

Table 2.1 summarizes the advantages and disadvantages of centralized, decentralized and hybrid approaches.

Although previous approaches did not address reallocation for incomplete tasks the approach described in this book considers reallocation of tasks, even if they were already assigned to agents in previous rounds. This reallocation increases the problem's complexity.

Table 2.1 Summary of task allocation approaches and their features

Feature/approach	Centralized	Decentralized	Hybrid
Handle unknown and dynamic situations	No	Yes	Partially
Communication and computation resources demand	Many	Few	Many
Scalability	No	Yes	Yes
Robustness	No	Yes	Yes
Flexibility	No	Yes	Yes
Agents team size	Small	Large	Medium
Need for a priori global information	Yes	No	No

References

1. Agmon N, Kaminka GA, Kraus S, Traub M (2010) Task reallocation in multi-robot formations. J Phys Agents 4(2):1–10
2. Attiya G, Hamam Y (2006) Task allocation for maximizing reliability of distributed systems: a simulated annealing approach. J Parallel Distrib Comput 66(10):1259–1266
3. Ball MG, Qela B, Wesolkowski S (2016) A review of the use of computational intelligence in the design of military surveillance networks. In: Recent advances in computational intelligence in defense and security. Springer International Publishing, pp 663–693
4. Barbagallo D, Di Nitto E, Dubois DJ, Mirandola R (2010) A bio-inspired algorithm for energy optimization in a self-organizing data center. In: Self-organizing architectures. Springer Berlin/Heidelberg, pp 127–151
5. Bayındır L (2016) A review of swarm robotics tasks. Neurocomputing 172:292–321
6. Berman S, Halász A, Kumar V, Pratt S (2007) Bio-inspired group behaviors for the deployment of a swarm of robots to multiple destinations. In: IEEE international conference on robotics and automation, pp 2318–2323
7. Blum C, Groß R (2015) Swarm intelligence in optimization and robotics. In: Springer handbook of computational intelligence, pp 1291–1309
8. Bonabeau E, Dorigo M, Theraulaz G (1999) Swarm intelligence: from natural to artificial systems. Oxford University Press Inc., New York, NY, USA
9. Brumitt B, Stentz A (1998) GRAMMPS: a generalized mission planner for multiple mobile robots. In: Proceedings of the IEEE international conference robotics and automation, Leuven, Belgium, vol 2, pp 1564–1571
10. Caloud P, Choi W, Latombe J, Le Pape C, Yim M (1990) Indoor automation with many mobile robots. In: Proceedings of the IEEE international workshop on intelligent robotics and systems (IROS), Ibaraki, Japan, vol 1, pp 67–72
11. Cao J, Yin B, Lu X, Kang Y, Chen X (2017) A modified artificial bee colony approach for the 0-1 knapsack problem. Appl Intell 1–14
12. Chen XW, Nof SY (2012) Conflict and error prevention and detection in complex networks. Automatica 48(5):770–778
13. Chu WW, Holloway LJ, Lan MT, Efe K (1980) Task allocation in distributed data processing. Computer 13(11):57–69
14. De Weerdt M, Zhang Y, Klos T (2007) Distributed task allocation in social networks. In: Proceedings of the 6th international joint conference on autonomous agents and multiagent systems, p 76
15. Dias MB, Zlot R, Kalra N, Stentz A (2006) Market-based multirobot coordination: a survey and analysis. Proc IEEE 94(7):1257–1270
16. Duan H, Li P (2014) Bio-inspired computation in unmanned aerial vehicles. Springer Berlin/Heidelberg, Berlin, Germany
17. Eberhart RC, Shi Y, Kennedy J (2001) Swarm intelligence. Elsevier
18. Fu B, Liang Y, Chen C (2015) Bio-inspired group modeling and analysis for intruder detection in mobile sensor/robotic networks. IEEE Trans Cybern 45:103–115
19. Gale D (1960) The theory of linear economic models. McGraw-Hill
20. Gerkey BP, Matarić MJ (2004) A formal analysis and taxonomy of task allocation in multi-robot systems. Int J Robot Res 23(9):939–954
21. Gerkey BP, Matarić MJ (2003) Multi-robot task allocation: analyzing the complexity and optimality of key architectures. In: IEEE international conference on robotics and automation, pp 3862–3868
22. Giordani S, Lujak M, Martinelli F (2010) A distributed algorithm for the multirobot task allocation problem. In: International conference on industrial, engineering and other applications of applied intelligent systems, pp 721–730
23. Groß R, Nouyan S, Bonani M, Mondada F, Dorigo M (2008) Division of labor in self-organized groups. In: Proceedings of the 10th international conference on simulation of adaptive behavior: from animals to animats. Springer-Verlag, Berlin, pp 426–436

24. Hussein A, Marín-Plaza P, García F, Armingol JM (2018) Hybrid optimization-based approach for multiple intelligent vehicles requests allocation. J Adv Transp
25. Jevtić A, Gutiérrez A, Andina D, Jamshidi M (2012) Distributed bees algorithm for task allocation in swarm of robots. IEEE Syst J 6(2):296–304
26. Jevtić A (2011) Swarm intelligence: novel tools for optimization, feature extraction, and multi-agent system modeling. PhD thesis
27. Jevtić A, Andina D, Jamshidi M (2014) Distributed task allocation in swarms of robots. In: Robotics: concepts, methodologies, tools, and applications. Information Science Reference, Hershey, PA, pp 450–473. https://doi.org/10.4018/978-1-4666-4607-0.ch023
28. Jones EG, Dias MB, Stentz A (2007) Learning-enhanced market-based task allocation for over-subscribed domains. In: Proceedings of the IEEE/RSJ international conference on intelligent robots and systems, San Diego, CA
29. Kalra N, Stentz A, Ferguson D (2005) Hoplites: a market framework for complex tight coordination in multi-agent teams. In: Proceedings of the international conference on robotics and automation (ICRA), New Orleans, USA, pp 1170–1177
30. Kao YH, Krishnamachari B, Ra MR, Bai F (2017) Hermes: latency optimal task assignment for resource-constrained mobile computing. IEEE Trans Mob Comput 16(11):3056–3069
31. Kapoor KN, Majumdar S, Nandy B (2015) Techniques for allocation of sensors in shared wireless sensor networks. J Netw 10(01):15–28
32. Kartik S, Murthy CSR (1997) Task allocation algorithms for maximizing reliability of distributed computing systems. IEEE Trans Comput 46(6):719–724
33. Khamis A, Hussein A, Elmogy A (2015) Multi-robot task allocation: a review of the state-of-the-art. In: Cooperative robots and sensor networks. Springer International Publishing, pp 31–51
34. Labella TH, Dorigo M, Deneubourg JL (2006) Division of labor in a group of robots inspired by ants' foraging behavior. ACM Trans Auton Adapt Syst (TAAS) 1(1):4–25
35. Lau HC, Zhang L (2003) Task allocation via multi-agent coalition formation: taxonomy, algorithms and complexity. In: Proceedings of the 15th IEEE international conference on tools with artificial intelligence, Sacramento, CA, USA, pp 346–350
36. Liu L, Michael N, Shell DA (2015) Communication constrained task allocation with optimized local task swaps. Auton Robots 39(3):429–444
37. Ma PR, Lee EY, Tsuchiya M (1982) A task allocation model for distributed computing systems. IEEE Trans Comput 31(1):41–47
38. Macarthur KS, Stranders R, Ramchurn SD, Jennings NR (2011) A distributed anytime algorithm for dynamic task allocation in multi-agent systems. In: Proceedings of the 25th conference on artificial intelligence, pp 701–706
39. Mataríc MJ, Sukhatme GS, Østergård EH (2003) Multi-robot task allocation in uncertain environments. Auton Robots 14(2):255–263
40. Nanjanath M, Gini M (2010) Repeated auctions for robust task execution by a robot team. Robot Auton Syst 58(7):900–909
41. Quijano N, Passino KM (2010) Honey bee social foraging algorithms for resource allocation: theory and application. Eng Appl Artif Intell 23(6):845–861
42. Ramchurn SD, Polukarov M, Farinelli A, Truong C, Jennings NR (2010a) Coalition formation with spatial and temporal constraints. In: Proceedings of the 9th international conference on autonomous agents and multiagent systems (AAMAS-10), Toronto, Canada, pp 1181–1188
43. Ramchurn SD, Farinelli A, Macarthur KS, Jennings, NR (2010b). Decentralized coordination in robocup rescue. Comput J 53(9):1447–1461
44. Robin C, Lacroix S (2015) Multi-robot target detection and tracking: taxonomy and survey. Auton Robots 1–32
45. Rowaihy H, Eswaran S, Johnson M, Verma D, Bar-Noy A, Brown T, Porta TL (2007) A survey of sensor selection schemes in wireless sensor networks. In: Proceedings of SPIE, vol 6562
46. Sankary N, Ostfeld A (2018) Multiobjective optimization of inline mobile and fixed wireless sensor networks under conditions of demand uncertainty. J Water Resour Plan Manag 144(8):04018043

47. Senanayake M, Senthooran I, Barca JC, Chung H, Kamruzzaman J, Murshed M (2016) Search and tracking algorithms for swarms of robots: a survey. Robot Auton Syst 75:422–434
48. Shehory O, Kraus S (1998) Methods for task allocation via agent coalition formation. Artif Intell 101(1):165–200
49. Sun Z, Liu Y, Tao L (2018) Attack localization task allocation in wireless sensor networks based on multi-objective binary particle swarm optimization. J Netw Comput Appl 112:29–40
50. Tan Y, Ding K (2016) Survey of GPU-based implementation of swarm intelligence algorithms. IEEE Trans Cybern 46:2028–2041
51. Tang Y (2016) Coordination of multi-agent systems under switching topologies via disturbance observer-based approach. Int J Syst Sci 1–8
52. Tindell KW, Burns A, Wellings AJ (1992) Allocating hard real-time tasks: an NP-hard problem made easy. Real-Time Syst 4(2):145–165
53. Turner J (2018) Distributed task allocation optimisation techniques. In: Proceedings of the 17th international conference on autonomous agents and multiagent systems, pp 1786–1787
54. Upadhyay D, Banerjee P (2016) An energy efficient proposed framework for time synchronization problem of wireless sensor network. In: Information systems design and intelligent applications. Springer India, pp 377–385
55. Vachtsevanos G, Tang L, Reinmann J (2004) An intelligent approach to coordinated control of multiple unmanned aerial vehicles. In: American helicopter society 60th annual forum, Baltimore
56. Walsh WE, Wellman MP (1998) A market protocol for decentralized task allocation. In: Proceedings of the international conference on multi-agent systems, pp 325–332
57. Wang J, Wang Y, Zhang D, Wang F, Xiong H, Chen C, Qiu Z (2018) Multi-task allocation in mobile crowd sensing with individual task quality assurance. IEEE Trans Mob Comput 17(9):2101–2113
58. Wichmann A, Korkmaz T, Tosun AS (2018) Robot control strategies for task allocation with connectivity constraints in wireless sensor and robot networks. IEEE Trans Mob Comput 17(6):1429–1441
59. Xue B, Zhang M, Browne W (2013) Particle swarm optimization for feature selection in classification: a multiobjective approach. IEEE Trans Cybern 43:1656–1671
60. Yan WQ (2016) Introduction to intelligent surveillance. Springer
61. Yuan Q, Guan Y, Hong B, Meng X (2013) Multi-robot task allocation using CNP combines with neural network. Neural Comput Appl 23(7–8):1909–1914
62. Zhan F, Wan X, Cheng Y, Ran B (2018) Methods for multi-type sensor allocations along a freeway corridor. IEEE Intell Transp Syst Mag 10(2):134–149
63. Zhang SZ, Lee CKM (2015) An improved artificial bee colony algorithm for the capacitated vehicle routing problem. In: Proceedings of the IEEE international conference on systems, man, and cybernetics (SMC), Kowloon, China, pp 2124–2128

Chapter 3
Multi-sensor Task Allocation Systems

The task allocation of a distributed heterogeneous multi-sensor system in dynamic and decentralized environments that have multiple tasks, with unknown a priori priorities that arrive at unknown locations at unknown times is a major challenge.

In most cases of distributed, dynamic, and decentralized environments it is reasonable to assume that each entity has only local or limited information and its own goals, which may or may not conflict with other entities' goals. In order to accomplish overall goals with such limited information and possibly conflicting local entity goals, it is inevitable that an effective control mechanism must be provided to coordinate and allocate tasks by exchanging information and decisions among participants.

This book focuses on robust scalable methods for sensor allocation with the ability to accommodate addition/subtraction of sensors during operation and handling problems existing in allocation methods, due to their inherent limitations, for task allocation with distributed heterogeneous sensors. In order to overcome such limitations, protocols need to be able to repeatedly identify the current state of the system and take proper actions to deal with allocation problems. Such protocols, which assume the responsibility of making decisions actively and triggering timely actions so that the overall system performance can be further improved, are defined as Task Administration Protocols (TAPs). The allocation problems handled with TAPs are:

1. Tasks that may have much higher priority over other tasks. This will occupy the sensors without the ability to perform other tasks that are close to their deadline and must be handled quickly before less urgent tasks.
2. Many tasks have the same priorities. The system may need to reprioritize them to avoid conflicts in sensor allocation.
3. Tasks with a low priority that require long execution times. These tasks may occupy the sensors.
4. Failure of a portion of sensors in the system that may affect the allocation process.

© Springer Nature Switzerland AG 2020
I. Tkach and Y. Edan, *Distributed Heterogeneous Multi Sensor Task Allocation Systems*,
Automation, Collaboration, & E-Services 7,
https://doi.org/10.1007/978-3-030-34735-2_3

3.1 Framework

A framework to enable efficient allocation of heterogeneous sensors to multiple tasks arriving at unknown different times and locations and to handle problems within the sensory system is described in this chapter. A dual-layer multi-sensor system was developed including: (1) a process layer—PL (in this research, multiple sensors and allocation algorithm HDBA for allocating them to efficiently complete tasks); (2) a monitoring layer—ML (in this research, task administration protocols for handling problems in the sensory system in the process layer). "TRAP" manages task priorities to allow better allocation of sensors to the most important tasks and to detect tasks when they occur. "ASAP" ensures that tasks will be treated as soon as possible and will not be unnecessarily delayed. "SRAP" allocates the best match of sensors to treat tasks and "STOP" was designed to ensure optimal availability of sensors in the system by time-out policy.

The book consists of three interrelated and independent parts that address multi-sensor task allocation:

1. Single-layer multi-sensor task allocation system—described in Chap. 5.
2. Dual-layer multi-sensor task allocation system—described in Chap. 7.
3. Fault tolerant multi sensor system with high availability—described in Chap. 9.

These interrelated parts are divided into specific modules necessary for efficient, scalable and robust task allocation. Figures 3.1, 3.2, 3.3 and 3.4 illustrate the overall architecture, the modules and the relevant book chapters where they defined.

Figure 3.2 illustrates comparison of state-of-the-art algorithms for several case studies.

Figure 3.3 illustrates a dual-layer system with TAPs, which can handle overloading, deception, and tampering.

Figure 3.4 illustrates the availability optimization of the sensor's operation to apply fault tolerance to the sensor network operation to make it robust enough to loss of sensors and/or the addition of new sensors.

3.2 Assumptions

The following assumptions were considered in the presented framework:

A1. Tasks are assigned to sensors, but their occurrence is unknown a priori.
A2. All of the tasks that are within a sensor's range can be allocated to that sensor.
A3. Decision-making about the allocation for each sensor takes place as soon as a new task is introduced.
A4. Sensors can be reallocated to another task during execution. An abandoned task keeps its remaining execution time, until a new sensor is allocated to it.

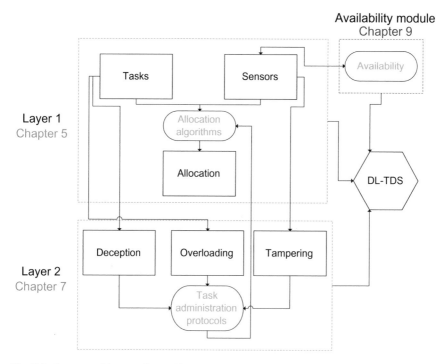

Fig. 3.1 System architecture for multi-sensor task allocation. The system consists of two layers and one availability module. The color corresponds to the book chapters

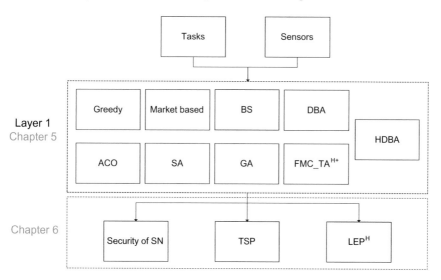

Fig. 3.2 Layer 1 with allocation algorithms architecture. The dashed boxes contain 9 allocation algorithms evaluated and compared for allocation (Sect. 5.3) and 3 evaluation scenarios for which the allocation algorithms were evaluated (Sects. 6.1, 6.2 and 6.3)

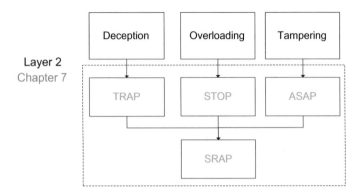

Fig. 3.3 Layer 2 with task administration protocols architecture. Green color indicates the task administration protocols that evaluated for the deception, overloading, and tampering problems (Sect. 7.3 and Chap. 8)

Fig. 3.4 Availability module architecture. System of type 1—sensors are set to perform without temporal restrictions. System of type 2—sensors with low MTBF values (optical sensors) are set aside and come into performance only when needed

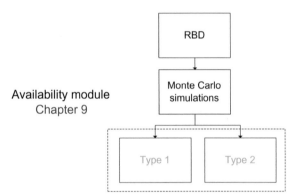

A5. Failure and repair distributions of the components are exponential (corresponding to the typical failure of electronic devices). Hence, failure and repair rates are constant.

A6. Upon failure, the sensor is repaired with normal time distributions of 240, 480, and 720 ± 2 h. These repair times include transporting the failed sensor to a depot, its repair time, and transport time back to the field.

A7. Sensor replacement in the field takes 5 h (modeled as insertion distribution). This time duration was set in order to add the impact of a typical replacement operation on the system's availability.

A8. Initial system deployment has no redundancies, in order to analyze the impact of adding redundant sensors to the system.

A9. Sensors collaborate with each other for task completion.

A10. Sensors use broadcast communication to share the estimated location and priority of the tasks.

Chapter 4
Evaluation Methodology

This chapter presents the analysis methods and the performance measures applied in this book. The described methods are evaluated by the use of extended examples in Chaps. 6 and 8. The evaluations include task attendance analyses that are conducted for both a single-layer and a dual-layer system to examine the effect and the performance of the described framework and algorithms.

4.1 Analyses

The aim of the analyses is to reveal the ability of the described methods to cope with the problems presented in Chap. 1 and specifically to: (a) analyze effective allocation and scalability in terms of number of tasks and sensors to be able to easily add or remove sensors to cope with an increasing number of tasks; (b) examine the performance of a dual-layer system, and its effect on optimizing task allocation in cases of sensor failure, conflicts in tasks priorities, and high time consuming tasks; and (c) optimize fault tolerance of the dual-layer task allocation system by Monte Carlo simulations.

The following analysis methods are applied in this book:

1. Algorithms for sensor allocation in a single-layer system—comparison of the state-of-the-art algorithms in terms of tasks completion times and the number of unallocated tasks.
2. Dual-layer system performance—comparison of the performance of a dual-layer system with task administration protocols (TAPs) and HDBA to a single-layer system with HDBA.
3. Availability optimization of the sensors' operation and comparison of the dual-layer system performance with optimized sensors' availability to a regular dual-layer system.

The system performance is evaluated by:

© Springer Nature Switzerland AG 2020

I. Tkach and Y. Edan, *Distributed Heterogeneous Multi Sensor Task Allocation Systems,*
Automation, Collaboration, & E-Services 7,
https://doi.org/10.1007/978-3-030-34735-2_4

1. Comparing to results of state-of-the-art algorithms and protocols using multi-agent simulation for the case studies of:

   (a) Security of supply networks.
   (b) A benchmark for result validation based on traveling sales man problem (TSP).
   (c) A Law Enforcement Problem (LEP).

2. Availability analysis (fault tolerance) of the multi sensor system using Monte Carlo simulation.

The numerical computations were implemented on a personal computer with 2.90 GHz CPU, and 12 GB of RAM, using Matlab R2015a, and JAVA SE8 programs to: (i) examine the performance of the algorithms for task allocation in a single-layer system by applying HDBA, DBA, BS, market-based, and greedy algorithms through multi-agent simulations; (ii) examine the scalability of HDBA in terms of number of tasks and sensors to be able to easily add or remove sensors to cope with increasing number of tasks; (iii) determine the influence of the bias parameters on HDBAs' performance adjusting the sensor swarm behavior; (iv) perform a multi-agent simulation analysis for result validation of HDBA based on a benchmark traveling salesman problem (TSP) in comparison to DBA, BS, ACO, GA, and greedy algorithms; (v) examine the performance of HDBA for police officers task allocation in a Law Enforcement Problem (LEP) in comparison to FMC_TA^{H+} and SA algorithms—all in Chap. 6; (vi) examine the performance of TAPs—TRAP, ASAP, SRAP, STOP in a dual-layer system, and their effect on optimizing task allocation in cases of sensor failure, conflicts in tasks priorities, and high time-consuming tasks; in Chap. 8; (vii) optimize fault tolerance of the dual-layer task allocation system by Monte Carlo simulations—in Chap. 9; and (viii) examine the performance of single and dual-layer systems through an analytical analysis—in Chap. 10.

4.2   Performance Measures

The following performance measures were analyzed, harnessing different algorithms and evaluations:

1. For the process layer, 100 independent runs of HDBA, DBA, BS, market-based, and greedy algorithms with 100 agents, were compared at the statistical confidence level of 95% for task allocation using the following performance measures:

   PM1. System performance.
   PM2. Tasks completion time.
   PM3. Number of unallocated tasks.
   PM4. Number of tasks allocated to sensors.

   and, three sensor distributions:

    1. Deterministic deployment—grid distribution.
    2. Random deployment—uniform distribution.
    3. Random and biased deployment—normal distribution.

2. In the TSP scenario evaluation, HDBA, DBA, BS, ACO, GA, SA, and greedy algorithms were compared for the performance measure of total distance in Berlin52 and A280 instances.

3. In the LEP scenario evaluation, 100 independent runs of HDBA, FMC_TA^{H+}, and SA algorithms were compared at the statistical confidence level of 95% for two case studies:

    CS1. Standard LEP with 25 agents to enable comparison with LEP benchmark.
    CS2. Extended LEP with 100 agents to evaluate swarm and to conduct a fair comparison with task allocation benchmark.

    The following performance measures were used:
    PM1. Team utility.
    PM2. Average execution delay.
    PM3. Percentage of abandoned tasks.
    PM4. Percentage of shared tasks.
    PM5. Average arrival time of agents to tasks.

4. For the monitoring layer, 100 independent runs of a dual-layer system using TAPs—TRAP, ASAP, SRAP, STOP was compared to a single-layer system using HDBA at the statistical confidence level of 95% for task allocation using the following performance measures:

    PM1. Number of treated tasks by each sensor for a different number of false tasks.
    PM2. Number of treated tasks by each sensor for a different number of high time-consuming tasks.
    PM3. Number of treated tasks by each sensor for a different number of failed sensors.
    PM4. Number of unallocated tasks for a different number of false tasks.
    PM5. Number of unallocated tasks for a different number of high time-consuming tasks.
    PM6. Number of unallocated tasks for a different number of failed sensors.
    PM7. Number of important tasks treated for a different number of false tasks.
    PM8. Number of important tasks treated for a different number of high time-consuming tasks.
    PM9. Number of important tasks treated for a different number of failed sensors.

5. For the fault tolerance analysis, 100 independent runs of a dual-layer system were analyzed at the statistical confidence level of 95% using the following performance measures:

    PM1. Percentage of availability for a type 1 system—sensors are set to perform without temporal restrictions.

PM2. Percentage of availability for a type 2 system—sensors with low MTBF values (optical sensors) are set aside and come into performance only when needed.

PM3. Number of processed tasks for a type 1 system.

PM4. Number of processed tasks for a type 2 system.

PM5. Influence of the number of redundant optical sensors on type 1 systems' availability with different repair times of 240, 480, and 720 h.

Chapter 5
Single-Layer Multi-sensor Task Allocation System

This chapter defines the multi–sensor task allocation in a single-layer system. The allocation problem is described and algorithms for multi-agent and multi-sensor task allocation are presented. Figure 5.1 illustrates the Layer 1 architecture of the system.

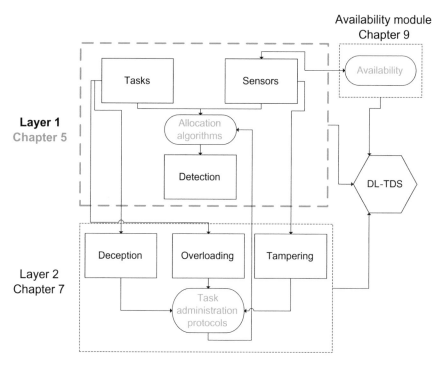

Fig. 5.1 Layer 1 in system architecture

© Springer Nature Switzerland AG 2020
I. Tkach and Y. Edan, *Distributed Heterogeneous Multi Sensor Task Allocation Systems*,
Automation, Collaboration, & E-Services 7,
https://doi.org/10.1007/978-3-030-34735-2_5

5.1 Definitions

The problem deals with real-time allocation of unpredictable, unknown tasks arriving at unknown times and locations. The task occurrence is dynamic and unpredictable with different levels of importance of each task and must be detected as fast as possible. Examples of such tasks include surveillance (gathering information on desired objects, [2, 72]), security monitoring (preventing theft of goods and threats, [50], fire monitoring (forest fire allocation and protection, [35, 84]), among many others. The sensors must be allocated to the tasks as fast as possible. The goal is to allocate to each sensor an appropriate task at an appropriate time (Fig. 5.2) and to ensure all tasks are completed in minimum time.

The system includes multiple sensors that are capable of performing each task with different performances (Fig. 5.3). Sensory performance is defined a priori based on the sensor's features, namely allocation distance, resolution, and response time. Each sensor can only be allocated to one task at any given time and can be reallocated to another task at any moment. The priority of a task is an application-specific scalar value, where a higher priority value represents a task that has higher importance and

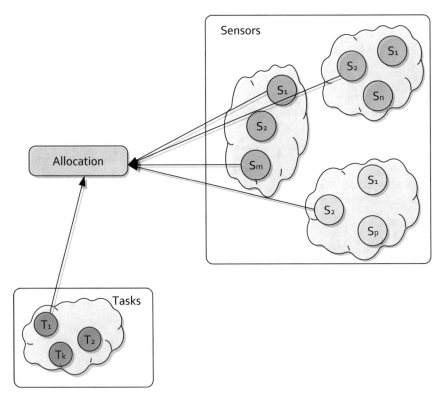

Fig. 5.2 Distributed multi-sensor system allocation scheme

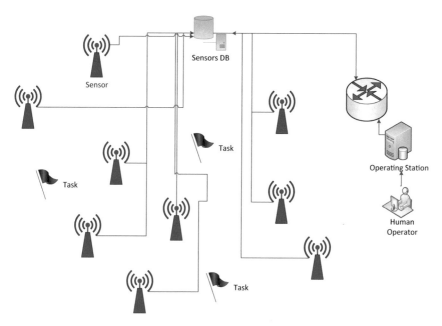

Fig. 5.3 A monitoring sensor network with sensors, tasks, and sensors DB that informs operators about the completed tasks, but does not contribute to the task allocation algorithm (after Tkach et al. [77])

must be attended to faster than other tasks. Higher priority tasks also have a higher benefit for completing them.

In this book, there is no limitation in the number of sensors that can be allocated to a single task. This eliminates the conflict of demanding the same task by multiple sensors. Once a sensor finds a task, it informs the neighboring sensors about the found task and its parameters, using broadcast communication (as in [23, 34, 38, 80]). This message is then forwarded by these sensors over the entire network, as in Ducatelle et al. [23]. From that moment, the sensors are aware of the detected task. Even though the sensors use broadcast communication to share the estimated location and priority of the tasks, task allocation is performed in a decentralized manner. Each sensor makes an autonomous decision that is based on the information that it has received [77].

When there are several tasks that require the same sensors, the allocation depends on the sensors' availability and performance, the physical distance of sensors from the tasks, and the priorities of individual tasks. The following assumptions were considered in the proposed scenario, similar to research performed in Jevtić et al. [39]:

A1. Tasks are assigned to sensors, but their occurrence is unknown a priori.
A2. All of the tasks that are within a sensor's range can be allocated to that sensor.

A3. Decision-making about the allocation for each sensor takes place as soon as a new task is introduced.
A4. Sensors can be reallocated to another task during execution. An abandoned task keeps its remaining execution time, until a new sensor is allocated to it.
A5. Sensors collaborate with each other for task completion.
A6. Sensors use broadcast communication to share the estimated location and priority of the tasks.

Following Tang and Ozguner [72], the system is defined with the following characteristics:

1. Sensors are stationary; and
2. Tasks remain stationary after their occurrence.

Based on the taxonomy proposed by Robin and Lacroix [67], the problem of allocating sensors to tasks corresponds to a localization and observation problem. The task localization involves several sensors and it is most often a multi-sensor problem to improve knowledge about the task involving selection of different viewpoints to maximize the information gain. The observation problem involves several sensors and several tasks in order to maximize the number of observed tasks and to minimize the time during which any task is not observed by at least one of the sensors.

A system objective function is developed to compute the expected value of system performance given parameters of the sensors and tasks. System performance is defined as the collective performance of sensors, task priorities values, and distances of sensors from tasks. Consider a population of N sensors to be allocated among M tasks. We denote the collective performance of the system by V_I, a nonnegative integer, calculated as:

$$V_I = \max\{S + \phi H\}, 0 < \phi < 1 \tag{5.1}$$

$$S = \sum_{i=1}^{M} \sum_{k=1}^{N} V_{ik} \cdot \frac{1}{D_{ik}} \tag{5.2}$$

$$H = \sum_{j=1}^{M} F_j, \; j \in \text{completed tasks} \tag{5.3}$$

$$V_{ik} = 0 \text{ if } k - \text{th sensor is not assigned to task i} \tag{5.4}$$

where S is the collective performance of the sensors, V_{ik} is the k-th sensor's performance on the i-th task, and N is the number of sensors in the system. H is the sum of the priorities of tasks in the system that were successfully completed, and φ is a bias parameter for the importance of H relative to S. F_j is the priority of j-th task and M is the total number of tasks. These values are pre-defined based on Neapolitan and Naimipour [57].

It is assumed that each sensor has a different performance based on the treated task. This value represents the heterogeneity of agents, by assigning each agent a different performance value based on its skills.

In example, for tasks occurring at night time, a night vision sensor will perform better than a regular CCD camera, but in daylight the CCD camera will have better performance. In a law enforcement problem, a traffic policeman will treat a traffic violation better than a murder detective, but will be incompetent to deal with a murder case.

Sensors are distributed in the arena with a particular distribution strategy and can be allocated to tasks within their allocation range. The sensors are stationary. One of the parameters that affects system performance is the distance of the sensors from tasks. The performance degrades as the distance increases, due to the degradation of sensors' recognition capabilities. The Euclidean distance between the sensor and the task in a two-dimensional arena is given by:

$$D_{ik} = \sqrt{(x_i - x_k)^2 + (y_i - y_k)^2} \tag{5.5}$$

where (x_i, y_i) and (x_k, y_k) represent task and sensor coordinates in the arena, respectively.

Let $F \in \{F_1, \ldots, F_M\}$ denote the set of normalized priorities of the available tasks in the queue. Normalized priorities are calculated as fractions of the sum of priorities of all available tasks:

$$F_i = \frac{f_i}{\sum_{j=1}^{M} f_j} \tag{5.6}$$

where f_i is a priority of task i.

In real-world scenarios, the priority is an estimated value that results from sensor readings or previously acquired knowledge. In a law enforcement problem, for example, the type of the task determines its priority. A noise complaint will have a lower priority than a robbery. A murder will have a higher priority than a robbery. In a supply network security problem, an unlocked door will have lower priority than missing cargo, but missing cargo will have lower priority than a fire in a central logistics center.

Each task has a time limit, or a deadline:

$$\Delta_i = \frac{1}{F_i} > 0, F_i > 0 \tag{5.7}$$

The deadline is a function of the priority, where Δ_i decreases with time until it reaches a value of 0, which indicates that the task i is not relevant anymore.

Once a sensor is allocated to a task, it monitors this particular task (with a performance V_{ik}). A task is considered to be detected once sensors are allocated to it.

Each task is assumed to have an initial completion time value t_{init}. Collaboration is achieved by sharing a task with multiple sensors. The task execution time value is modeled as a function of the initial completion time that is required for the task and the performances of coordinating sensors that are allocated to it. This dynamic value is continuously updated Eq. (5.8). The higher the cumulative performance of the coordinating sensors, the lower the execution time needed for the shared task.

$$
t_{ex_i} = \begin{cases} t_{init} \frac{1}{\sum_{k=1}^{M} V_{ik}} - t_{p_i} & \text{if } \sum_{k=1}^{M} V_{ik} > 0, \Delta_i > 0 \\ t_{init} & \text{if } \sum_{k=1}^{M} V_{ik} = 0, \Delta_i > 0 \end{cases} \tag{5.8}
$$

where t_{ex_i} is the remaining execution time of task i and t_{p_i} is the elapsed time of task i execution. When t_{ex_i} reaches 0 value, the task is completed, and it is removed from the arena by the broadcast communication.

When new tasks are announced, steps (1–8) are updated respectively. The overall task completion time T is defined as a sum of the individual tasks completion times t_{ci}:

$$
T = \sum_{i=1}^{M} t_{c_i} \tag{5.9}
$$

$$
t_{c_i} = t_{i_{f\_completion}} - t_{i_{arrival}}, \quad \text{if } \Delta_i > 0 \tag{5.10}
$$

The task completion time is derived by the amount of time elapsed from its arrival $t_{iarrival}$ to the full completion $t_{if\_completion}$.

5.2   Algorithms for Multi-agent Task Allocation

5.2.1   Meta Heuristics

Meta heuristics usually have an iterative behavior and have emerged in recent years as successful alternatives to classical approaches in solving complex optimization problems in a reasonable time [8, 12, 71]. Meta heuristic algorithms include swarm intelligence, evolutionary computation, simulated annealing, tabu search, and hill climbing among others [14, 58]. Evolutionary computation includes optimization algorithms that are inspired by Darwinian evolution. In these algorithms, a set of solutions is called a population. Each iteration is used to create a new generation from the current population. This approach typically has two operators, called recombination or crossover, which recombines two or more individuals and mutations, which modifies a single individual [29]. Based on the objective function, individuals from the current generation are selected for recombination and mutation, or to be repeated for the next generation without any change, while individuals with higher fitness have a higher probability to be selected. Evolutionary computation algorithms include evolutionary programming [26], evolutionary strategies [66], and genetic algorithms [36].

Tabu Search, introduced by Glover [32], uses search history in order to explore the solution space. This algorithm applies the best improvement local search with memory to avoid local optima. The memory is implemented as a tabu list that keeps track of the most recently visited solutions and forbids moves toward them. At each iteration, the best solution is chosen from the allowed set as the new current solution.

Hill climbing is a local search optimization algorithm [60, 63, 69]. It is an iterative algorithm that attempts to find a better solution by making changes to the current solution. If the change produces a better solution it is selected to be further modified until the best solution is obtained [69]. Hill climbing can often produce a better result when the amount of time available to perform a search is limited. It is good for finding local optima, but it is not necessarily guaranteed to find the optimal solution.

Simulated annealing introduced in the area of combinatorial optimization by Kirkpatrick et al. [47], is a state-of-the-art meta heuristic for approximating global optimization in a large domain [69] and one of the first having a strategy for escaping local optima. It simulates a physical annealing process in metallurgy, where particles of a solid arrange themselves into a thermal equilibrium, a technique involving heating and controlled cooling of a material to increase the size of its crystals and reduce their defects. Simulated annealing interprets slow cooling as low decrease in the probability of accepting worse solutions as it explores the solution space. Accepting worse solutions is a fundamental property of metaheuristics because it allows for a more extensive search for the optimal solution (explores the solution space), reducing the chances of being converged to local optima.

Heuristic methods combine intelligently different concepts for exploring the search space in order to find near-optimal solutions [59] and to enable escaping from local optima. However, they do not always guarantee to find a globally optimal solution [14].

5.2.2 Swarm Intelligence

Swarm-based systems are typically made of a population of simple agents (system entities that make decisions about their local environment) interacting locally with one another and with their environment to achieve a common goal [40]. These approaches simulate the behaviors of insects and animals to assign tasks to agents (Fig. 5.4). Swarm intelligence methods [21] are mainly used for agent systems in unknown environments and optimization for task allocation, clustering, navigation, and path formation [11, 68]. Since the group collaboration among individuals is distributed, an individual fault does not affect the global solution [37, 46]. Computational Swarm Intelligence represents the group of Computational Intelligence algorithms that model swarming behaviors in nature. Computational Intelligence belongs to the broader field of Artificial Intelligence, and of the paradigms that relate to some kind

Fig. 5.4 Swarm intelligence in nature. School of fish and swarms of bees and ants. Photos by Karen Kayser, Topcools Tee and Mikhail Vasilyev on Unsplash

of biological or naturally occurring system [5]. Swarm intelligence methods have high robustness, scalability, and are suitable for distributed multi-agent systems [9]. The benefits of collaboration can be significant in situations where global knowledge of the environment does not exist. Agents within the group interact by exchanging locally available information such that the global objective can be obtained more efficiently than it would be done by a single agent. The group of agents acting in such a manner can be referred to as a swarm. The problem-solving behavior that emerges from the interactions of such agents is called Swarm Intelligence. Algorithmic models of such behavior are referred to as Computational Swarm Intelligence [25].

Examples of collective behavior in nature are numerous. They are based on a direct or indirect exchange of information about the environment between the members of the swarm [24]. Although the rules governing the interactions at the local level are usually easy to describe, the result of such behavior is difficult to predict [53]. However, through collaboration, the swarms in nature are able to solve complex problems that are crucial for survival in a dynamically changing environment [15, 74].

There are many examples of swarming behavior in the literature as it has always intrigued scientists for its better understanding. Ants communicate by laying pheromone on their route in order to find the shortest path to a food source by using foraging behavior [11, 21, 28]. Termites cooperate by stigmergy to build complex nest structures without having any global knowledge of the environment [28, 40]. Bees use dancing to recruit other members in the swarm to follow them to the location in the field that is rich with nectar [13]. Birds gather in flocks and fish form schools to have better chances of survival against predators [62]. Bacteria locally exchange information about their environment, and so forth.

Pham et al. [61] developed the Bees Algorithm (BA), which in its basic version performs a random search combined with a neighborhood search and can be used for optimization. The algorithm exploits the concept of the central dance floor in order to select the fittest sites, but no direct communication between the swarm members exists. The recruitment of the bees can be done in a deterministic way according to the fitness values associated with the sites, or these fitness values can be used to determine the probability of the bees being selected. Together with scouting, this differential recruitment is a key operation of the BA. The algorithm is performed in iterations, and it is stopped when the solution is found within the provided error margin or when the maximal number of iterations is reached. Bailis et al. [7] proposed a model of the bee colony foraging to investigate the value of sharing food source position information in different environments. The authors show through simulations that in environments of highly-scattered food, relying solely on private information about previously encountered food sources is more efficient than sharing information. However, in the nectar-rich environments, it leads to decreased foraging efficiency.

Karaboga and Akay [45] proposed a bee colony-based optimization algorithm called Artificial Bee Colony (ABC). The concept of the central dance floor is applied through different roles that bees have in the swarm, namely scout, onlooker, and employed bees. The algorithm uses recruitment based on the fitness values of the food sources and applies neighborhood search for solution improvements. A scout bee is randomly sent to search for a new food source when a previously found source is abandoned. Jevtić [40] proposed a distributed bees algorithm for task allocation in a swarm of robots. This algorithm assigns the robots in a swarm to the found tasks in such a way that the final distribution is proportional to the tasks' qualities. The robots were designed to use broadcast communication to inform other robots in the range about the estimated location and the quality of the found task.

Few other algorithms inspired by bees' behavior appeared in literature, such as BeeHive [79], BCO [75], Virtual Bee Algorithm [82], HBMO [1], etc., and they have mostly been applied to solving combinatorial optimization problems.

This book will show how to extend the use of swarm intelligence algorithms to the problem of allocating heterogeneous static agents to dynamic tasks with a dynamic reallocation of agents to ensure scalability and robustness.

5.2.3 Market-Based Approaches

Market-based approaches have gained popularity in recent years, mainly because they offer a good compromise between finding a good allocation and the communication speed [46]. Economic problems have involved humans throughout history to pursue individual profit. This pursuit led to the formalism of the principles of the market economy. These principles have inspired computational market-based approaches that are applied to multi-agent allocation [41]. The first method of applying market-based principles was Contract Net Protocol or CNP [70]. It allocates tasks to resources through the negotiation of contracts. It establishes a virtual economy with agents as buyers, tasks as products, and virtual money as currency.

Market-based approaches have been applied to resource allocation on computers [10, 48, 81], multi-agent [6, 42], and multi-robot coordination [18, 49, 55, 85].

In market-based systems, agents act to maximize their individual profit and simultaneously improve the efficiency of the team. They are paid based on their bid for tasks they complete and must pay the value of the resources they consume. The bids are managed by the auctioneer. The auctioneer can be a centralized supervisor agent or one of the agents. In each auction, agents compete by bidding to be allocated to tasks and to maximize their profit. The new allocation achieved by bidding can result in better solutions. One of the first distributed market-based algorithms was M+ [17], proposed for the cooperation among multiple robots [16]. In this algorithm, robots employ a two-step principle to calculate the cost of a task considering the next one in order to increase the efficiency of the solution. In MURDOCH [31, 30] robots do not participate in auctions during task execution, and tasks are only allocated to idle robots. It was reported to result in less efficient solutions than M+ because it does not consider future tasks and uses a purely greedy approach. However, the advantage of MURDOCH is that it uses less communication. TraderBots [19] is a market-based system designed to work in dynamic environments. This system applies scheduling to allocate multiple tasks to each robot and allows bidding on several tasks simultaneously. It was reported to make efficient allocations despite partial failures of the robots and communications (Dias et al. [20]).

Ramchurn et al. [65] present a coalition formation with spatial and temporal constraints problem (CFSTP). Agents form coalitions to jointly work on tasks with spatial constraints and deadlines. Agents have hard deadlines and are sharing tasks at non-additive rates to maximize the number of tasks completed. CFSTP was originally solved using Coalition Formation with Look-Ahead (CFLA) [65] that applies two heuristics: allocating the smallest possible coalition and maximizing the number of other tasks that can be completed before the deadline in the next time step. A distributed Max-sum DCOP algorithm was used to solve CFLA [52, 64, 83]. Walsh and Wellman [78] presented a mechanism to accomplish a similar goal via a distributed market protocol for allocating tasks to agents that contend for scarce resources. Agents trade tasks at prices determined by an auction protocol. In Jones et al. [42], a market-based mechanism was applied to a real-world firefighting problem. In that work, fire incidents defined as tasks were assigned to firefighting units according to the order in which the fires were discovered; each fire had an auctioneer, and agents computed the change in utility for being allocated a task, using either an optimal or near-optimal ordering, and submitted this change as a bid. The auctioneer greedily and irrevocably allocated the task to the agent with the highest bid. Once a fire was assigned to an agent, the agent was responsible for it until it was extinguished.

Several market-based works focused on task allocation problems that have restrictions between tasks. Hoplites [43, 44] is a market-based system that applies both passive and active coordination. Tasks are completed faster with passive coordination, resulting in less communication usage. Active coordination is used to enable agents to actively affect the bidding of other agents. Market-based algorithms for robot routing in open spaces are presented in Mosteo et al. [54]. Guerrero and Oliver [33] presented market approaches to find an optimal number of robots needed to solve specific tasks. Nanjanath and Gini [56] presented a variation of CNP for solving task allocation problems. Their system allowed reallocation of tasks to cope with dynamic environments. Zlot and Stentz [85] addressed the market-based allocation of complex tasks with task decomposition and allocation phases considered together instead of separated as usual.

Market-based approaches capture the respective strengths of both distributed and centralized approaches [46]. They can distribute much of the planning and execution over the team and thereby retain the benefits of distributed approaches, including robustness, flexibility, and speed. They also have elements of centralized systems to produce better solutions: auctions concisely gather information about the team and distribute resources in a team-aware context. However, if the communication costs are too high in the task allocation process, once there are failures in agent's communication, the performance will degrade noticeably [73], so these methods fit for small and medium-scale task allocation.

The task allocation problem can be solved by various algorithms and meta heuristics. In this book, several of them were implemented and compared, including HDBA algorithm, simulated annealing, ant colony optimization, bees system, Fisher-market, and genetic algorithm.

5.3   Specific Algorithms for Multi-sensor Task Allocation

This section describes nine algorithms suitable for multi-sensor task allocation: HDBA, DBA, Greedy algorithm, Market-based algorithm, Bee System (BS), Fisher Market Clearing for Task Allocation with Heterogeneous Agent (FMC_TA^{H+}), Ant Colony Optimization (ACO), Genetic algorithm (GA), Simulated Annealing (SA).

5.3.1   Distributed Bees Algorithm

In this algorithm, each sensor is represented as a 'bee', and sensor utility, p_{ik}, is defined as a probability that the sensor k is allocated to the task i and depends on both priority and the distance of the task/task from the sensor:

$$p_{ik} = \frac{F_i^\alpha \left(\frac{1}{D_{ik}}\right)^\beta}{\sum_{j=1}^{M} F_j^\alpha \left(\frac{1}{D_{jk}}\right)^\beta} \quad \text{if } \Delta_i > 0 \qquad (5.11)$$

where α and β are control parameters that bias importance of the priority and distance, respectively ($\alpha, \beta > 0$; $\alpha, \beta \in R$). The probabilities p_{ik} are normalized, and it is easy to show that:

$$\sum_{i=1}^{M} p_{ik} = 1 \qquad (5.12)$$

The DBA decision-making mechanism uses a wheel-selection rule, where each sensor has a probability with which it is allocated to the task from a set of available

Fig. 5.5 The waggle dance of a bee. The bee performs a waggle dance to indicate the direction, quality and distance of the food source. Photo by Damien Tupinier on Unsplash

tasks. Once all the sensors' utilities are calculated, each of them selects a task by "spinning the wheel".

The information is then exchanged by the sensors using broadcast communication, inspired by the waggle dance of bees in nature [27] which is used to share with other bees the direction, quality and distance of the found source food (Fig. 5.5).

5.3.2 Heterogeneous Distributed Bees Algorithm

Researchers have proposed a distributed bees algorithm that is suitable for implementation in a multi-agent system where cooperating agents interact by exchanging locally available information, such that the global objective is obtained more efficiently than it would be done by agents that perform tasks individually. This swarm intelligence algorithm was enhanced to control a heterogeneous sensor network allocation.

The DBA function of sensors' utility (HDBA) was modified for the case of heterogeneous sensors with different performances. This modification is assumed to improve system performance as it can correlate the sensors' utility function with the value of their performances.

In order to define HDBA for heterogeneous sensors, a new task's utility value (V_{ik}) is defined as a function of the sensor's performance on that task. When a sensor receives information about the available tasks it calculates its performance for that task. The sensor's utility function is updated accordingly, and depends on the task's priority, the distance from the task, and the sensor's performance on that task:

$$p_{ik} = \frac{F_i^\alpha \left(\frac{1}{D_{ik}}\right)^\beta V_{ik}^\gamma}{\sum_{j=1}^{M} F_j^\alpha \left(\frac{1}{D_{jk}}\right)^\beta V_{jk}^\gamma} \quad \text{if } \Delta_i > 0 \tag{5.13}$$

where γ is a control parameter that biases the importance of the sensor's performance and V_{ik} is the performance of sensor k on task i.

The HDBA decision-making mechanism applies the same wheel-selection rule used in DBA to choose from a set of available tasks.

Tasks arriving at different times trigger a new decision-making iteration of the algorithm.

Figures 5.6 and 5.7 illustrate the allocation process.

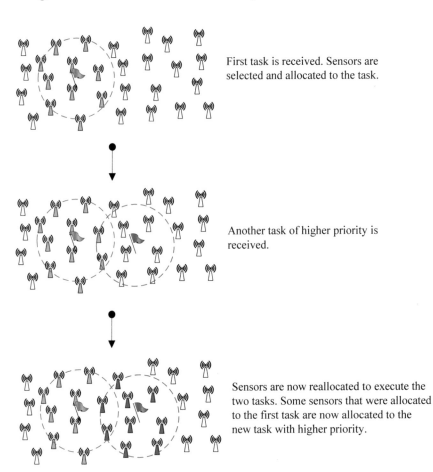

First task is received. Sensors are selected and allocated to the task.

Another task of higher priority is received.

Sensors are now reallocated to execute the two tasks. Some sensors that were allocated to the first task are now allocated to the new task with higher priority.

Fig. 5.6 Sensors to tasks allocation process. The sensors' range coverage was set to equal values for illustration purposes (after Tkach et al. [76])

Fig. 5.7 Allocation
probabilities for each task
(after Tkach et al. [76])

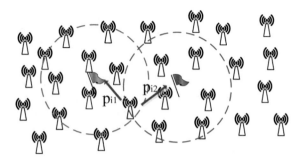

   In Fig. 5.7 two different tasks arrived at different times. The normalized priority
of task 1 is given by F_1 and of task 2 by F_2. The Euclidian distance of the sensors
from task 1 and task 2 is D_{1k} and D_{2k}, respectively. The performance of the sensors
on task 1 and task 2 is V_{1k} and V_{2k}, respectively. The probabilities of the i'th sensor
to be allocated to task 1 and task 2 are p_{i1} and p_{i2}, respectively. Upon the arrival of
the first task, HDBA is executed to calculate the probabilities of sensors allocation.
Since there are no other tasks, sensors are allocated to the task that is within their
allocation range, and they cooperate to complete this task. Upon the arrival of the
second task, another decision-making iteration of HDBA is triggered to calculate the
new probabilities of sensors allocation. Based on the calculated probabilities, HDBA
converges to a new allocation. In this allocation, additional sensors are allocated to
the second task and some sensors that were allocated to the first task are reallocated
to the second task, such that the overall task completion time is minimized. HDBA
is executed until the termination condition is met.
   A numerical example of HDBA's allocation process for a simplified problem is
described in Chap. 10.
   The DBA/HDBA, similar to other swarm intelligence algorithms, include a population of simple agents interacting locally with one another and with their environment. Agents make an iterative construction of solutions to search the solution
space considering previous solutions to optimize their search. This is a probabilistic multi-agent algorithm that uses a probability distribution to make the transition
between each iteration. The probabilistic nature of this algorithm allows it to avoid
local optimum by choosing sometimes worse solutions using the wheel-selection
rule. Accepting worse solutions is a fundamental property of metaheuristics because
it allows for a more extensive search for the global optimal solution.

Pseudocode of the HDBA algorithm:

1. Initialize:
   Set $t=0$
   Place N bees on sensors
2. If *termination condition met* then go to step 5
   else
      Upon task arrival calculate the new task priority F_i
      Calculate distances of sensors from task D_{ik}
      Calculate performances of sensors on tasks V_{ik}
      Go to step 3
3. If *new task arrived* then go to step 2
   For $i=1$ to M do
      For $k=1$ to N do
         Calculate probabilities for each sensor p_{ik}
      Apply wheel-selection rule
   Allocate sensors according to the selection
4. If $(t_{exi}=0)$ then set $\Delta_i=0$ and go to step 2
   else
      Go to step 3
5. Finalize:
   Calculate V_I
   Calculate T
   Stop

5.3.3 Market-Based Algorithm

A market-based algorithm that uses bids from agents in auctions for distributed sensing tasks is presented. Agents compete to win tasks by placing bids in auctions to produce efficient allocations. Each distributed agent computes a cost for completing a task and broadcasts the bid for that task. The auctioneer agent decides the best available bid, and the winning bidder attempts to perform the task (Fig. 5.8). In this algorithm, the bid of sensor k to the task i is defined as (5.14).

$$\text{Bid}_{ik} = F_{ik} + \delta \left(V_{ik} \frac{1}{D_{ik}} - F_{ik} \right) \text{ if } \Delta_i > 0 \qquad (5.14)$$

where F_{ik} serves as the reservation price of task i, and δ is a control parameter with values between 0 and 1. Sensor k is allocated to a task if it maximizes Eq. (5.15).

$$\text{Select} = \max(\text{Bid}_{ik}) \qquad (5.15)$$

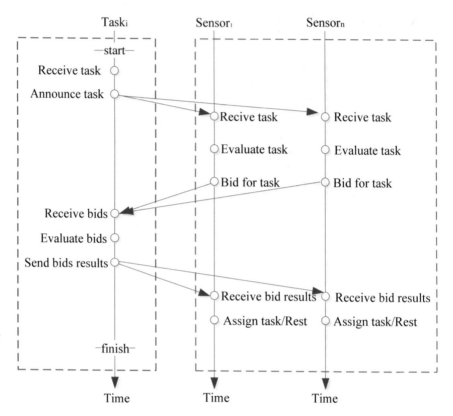

Fig. 5.8 Workflow diagram for the Market-based algorithm (after Tkach et al. [76])

5.3.4   Greedy Algorithm

A greedy algorithm for a multi-task observation problem with broadcast messaging is presented. This algorithm is set to perform sensor allocation based on the best possible allocation of each individual sensor to a task that maximizes V_{ik}/D_{ik}, where V_{ik} is the k-th sensor's performance on the i-th task and D_{ik} is the Euclidean distance between the sensor and the task.

$$task_i = \max_{i \in Z}\left(\frac{V_{ik}}{D_{ik}}\right) \text{ if } \Delta_i > 0 \tag{5.16}$$

where $task_i$ is the chosen task of the k-th sensor out of all the possible tasks for allocation, and Z is the group of tasks in k-th sensor range out of M available tasks.

5.3.5   Bee System

A Bee System (BS) used in Lučić and Teodorovic [51] for a traveling salesman problem is described. This algorithm is based on the foraging behavior of an artificial colony of bees to optimally allocate sensors to tasks. When a new task is introduced, the algorithm is able to replace the obtained allocation with a new improved allocation.

$$p_{ik} = \frac{e^{\alpha F_i - \beta D_{ik}}}{\sum_{j=1}^{M} e^{\alpha F_j - \beta D_{jk}}} \text{ if } \Delta_i > 0 \qquad (5.17)$$

where α and β are control parameters that bias importance of the priority and distance, respectively ($\alpha, \beta > 0$; $\alpha, \beta \in R$).

5.3.6   Fisher Market Clearing Task Allocation

A Fisher Market Clearing for Task Allocation with Heterogeneous Agent algorithm (FMC_TA^{H+}) computes an allocation by modeling agents as buyers, tasks as goods, and utilities as preferences [3]. Individual tasks, especially important ones, may be divided between multiple agents who share the workload. FMC_TA^{H+} schedules the allocation under inter-task and inter-agent constraints. The tasks allocated to each agent are ordered by importance. Task sequences are then modified based on inter-agent constraints. Specifically, because shared tasks cannot be started until all assigned agents have arrived, agents do not benefit from arriving earlier than the latest agent. For each agent scheduled to arrive early to a shared task, the next task is scheduled before the shared task if this move does not delay the execution of the shared task. The stages of the FMC_TA^{H+} algorithm are:

1. Generate a Fisher market instance;
2. Find a Fisher market clearing solution and acquire an allocation of tasks to agents;
3. Order the tasks allocated to each agent.

Each task Ψ_i in FMC_TA^{H+} is represented as k subtasks Ψ_{is}, where k is the number of agents with different skills in LEP. The matrix U is a 3-dimensional matrix of size $n \times m \times k$ to be used as an input for the Fisher Market Clearing mechanism. The utility function is:

$$u_{ik}(x_{ik}) = (h_{ik} x_{ik})^{\Gamma_i} \qquad (5.18)$$

where $u \in U$, $\Gamma_i \in (0, 1)$ and $h_{ik} = r_{iks}^{1/\Gamma_i}$. x_{ik} is the amount of task i that is allocated to agent k.

Each entry r_{iks} at time t represents the personal utility of agent k when immediately moving to handle subtask Ψ_{is}. If agent k does not possess skill s or task Ψ_i does not require it, the value of entry r_{iks} is zero. The utility for performing task Ψ_{is} depends on the number of agents that work simultaneously on that task and is denoted by the non-negative capability function $Cap(\Psi_{is}, q)$. This can represent the minimum required or maximum allowed number of agents by setting capability to 0 for fewer agents or by not increasing the value of $Cap(\Psi_{is}, q)$ when more than the maximum number of required agents cooperate on a task. The utility is constructed by optimistically ignoring the inter-task ordering constraints and assuming the maximum value for the capability function. A market clearing solution is a price vector Φ specifying a price Φ_{ik} for each good (task) i that allows each buyer (agent) k to spend all his money on goods that maximize bang-per-buck (r_{iks}/Φ_{ik}) while all goods in the market are sold (Amador et al. [4]).

$$r_{iks} = \Delta(\Psi_{is}, t) \max_{\vec{q}} \{Cap(\Psi_{is}, \vec{q})\} - \pi(CT_k, \Delta w) \tag{5.19}$$

where $\Delta(\Psi, t)$ is the soft deadline function, $\vec{q}$ is the number of agents performing a task, π is the penalty for task interruption, Δw is the amount of work completed when the task is interrupted, Cap is the capability function, and CT_k is the current task being performed by agent k.

5.3.7 Ant Colony Optimization

An ant colony swarm algorithm called Ant Colony Optimization (ACO) was developed by Dorigo et al. [22]. In this method, each ant is an agent that chooses sensors with a probability that is a function of the sensor's performance and the amount of trail laid. To force the ant to make legal selections, transitions to already visited sensors are disallowed until a cycle is completed (this is controlled by a *tabu list*); when it completes a cycle, it lays a substance called trail on each sensor visited (Eq. 5.20).

$$\upsilon_i(t + n) = \rho \times \upsilon_i(t) + \Delta \upsilon_i, \text{ if } \upsilon_i \geq \upsilon_0 \tag{5.20}$$

where υ_i is the intensity of ant's i trail at time t. n is the number of sensors (every n iterations – cycle, each ant completed a cycle). ρ is a coefficient such that $(1 - \rho)$ represents the evaporation of trail between time t and $t + n$, υ_o is a threshold.

$$\Delta \upsilon_i = \sum_{k=1}^{m} \Delta \upsilon_i^k \tag{5.21}$$

where $\Delta \upsilon_i$ is the quantity per unit of trail laid on sensor i by the k-th ant between time t and $t + n$; it is given by:

$$\Delta v_i^k = \begin{cases} \frac{Q}{L_k} & \text{if k-th ant uses path i in its cycle} \\ 0 & \text{otherwise} \end{cases} \qquad (5.22)$$

where Q is a constant and L_k is the tour length of the k-th ant. *Tabu list* saves the sensors already visited up to time t and forbids the ant from visiting them again. When a tour is completed, the *tabu list* is used to compute the ant's current solution (i.e., the set of sensors being reached).

The probability of going to sensor i for the k-th ant is given by:

$$p_i^k(t) = \begin{cases} \dfrac{v_i^{\psi}(t) \times \xi_i^{\omega}}{\sum_{k \in allowed_k} v_i^{\psi}(t) \times \xi_i^{\omega}} & \text{if } i \in allowed_k \\ 0 & \text{otherwise} \end{cases} \qquad (5.23)$$

where visibility ξ_i is the collective performance of the selected sensors S, $allowed_k$ is the set of sensors not in *tabu list*, ψ and ω are parameters that control the relative importance of trail versus visibility.

The stop criterion is given by a threshold that quantifies the desired performance of the sensors for a given task. The optimal sensor allocation is given by:

$$V_{OS} = \sum_{k \in tabu \text{ list}} S_j \geq threshold \qquad (5.24)$$

where j is the index of sensors in the *tabu list*.

Pseudocode of the ACO algorithm

1. Initialize:
   Set $t=0$
   $\Delta v_i^k = 0$
   Place N ants on sensors
2. If *end of mission* then go to step 5
   else
       Clear *tabu list*
       Upon task arrival calculate the new task priority f_i
       Update queue of tasks based on their priorities
       Go to step 3
3. While $\Delta_i > 0$:
       If *new task arrived* then go to step 2
           For $k=1$ to N do
           Choose the sensor i with probability $p_i^k(t)$
           Insert sensor i to *tabu list*
4. If $(t_{ex_i}=0)$ then set $\Delta_i=0$ and go to step 2
   else
       Go to step 4
5. Finalize:
       Calculate V_I
       Calculate T
       Stop

5.3.8   Genetic Algorithm

A genetic algorithm (GA) is an evolutionary algorithm that is used to solve optimiza-tion problems with non-polynomial complexity. The evolution usually starts from a population of randomly generated individuals consisting of legitimate candidate solutions, and is an iterative process, with the population in each iteration called a generation. In each generation, the fitness of every individual in the population is evaluated; the fitness is usually the value of the objective function in the optimiza-tion problem being solved. The more fit individuals are stochastically selected from the current population, and each individual's genome is modified by crossover and mutation to form a new generation. The new generation of candidate solutions is then used in the next iteration of the algorithm. Commonly, the algorithm terminates when either a maximum number of generations has been produced, or a satisfactory level of the objective function has been reached.

Pseudocode of GA

Initialize population
        Generate random genotypes representing legitimate solutions of sensor to tasks
allocation
        While not terminated do
                For each chromosome from population
                        Compute fitness functions V_l
                Make next population
                        Select parents
                        Apply crossover by recombining pairs of parents
                        Apply mutation to offspring by changing an allocation in a
                genotype
                        Evaluate new solutions by computing fitness functions V_l
                        Store the chromosome that obtains the best fitness function
        End while
Take the stored chromosome

5.3.9   Simulated Annealing

Simulated Annealing (SA) is a metaheuristic inspired by the physical annealing in metallurgy. It uses the objective function that implicitly maximizes the team utility. SA explores the solution space using random transitions to generate new solutions, and redusing the temperature value to converge. SA allows occasional transitions leading to worse solutions, but it helps keep the search from getting stuck in local optima.

Pseudocode of SA

```
Create initial solution Sol
Initialize temperature t
repeat
    for i = 1 to iteration-length do
        Generate a random transition from Sol to Sol_i
        if Cost(Sol) ≥ Cost(Sol_i) then
            Sol = Sol_i
        else if e^(Cost(Sol)−Cost(Soli))/(current temperature) > random [0,1) then
            Sol = Sol_i
        end if
    end for
    Reduce temperature t
until no change in Cost(Sol)
return Sol
```

References

1. Afshar AB, Haddad O, Mariño MA, Adams BJ (2007) Honey-bee mating optimization (HBMO) algorithm for optimal reservoir operation. J Franklin Inst 344(5):452–462
2. Altahir AA, Asirvadam VS, Hamid NHB, Sebastian P, Saad NB, Ibrahim RB, Dass SC (2018) Optimizing visual sensor coverage overlaps for multiview surveillance systems. IEEE Sens J 18(11):4544–4552
3. Amador S, Zivan R (2017) Incentivizing cooperation between heterogeneous agents in dynamic task allocation. In: Proceedings of 16th international conference on autonomous agents and multiagent systems (AAMAS), São Paulo, Brazil, pp 1082–1090
4. Amador S, Okamoto S, Zivan R (2014) Dynamic multi-agent task allocation with spatial and temporal constraints. Proceedings of the 2014 international conference on Autonomous agents and multi-agent systems,1495–1496
5. Andina D, Pham DT (2007) Computational intelligence: for engineering and manufacturing. Springer-Verlag, New York Inc
6. Baert Q, Caron AC, Morge M, Routier JC (2018) Fair multi-agent task allocation for large datasets analysis. Knowl Inf Syst 54(3):591–615
7. Bailis P, Nagpal R, Werfel J (2010) Positional communication and private information in honeybee foraging models. In: Dorigo M, Birattari M, Di Caro G, Doursat R, Engelbrecht A, Floreano D, Gambardella L, Grob R, Sahin E, Sayama H, StÄutzle T (eds) Swarm intelligence, volume 6234 of lecture notes in computer science, Springer Berlin, Heidelberg, pp 263–274
8. Balamurugan R, Natarajan AM, Premalatha K (2015) Stellar-mass black hole optimization for biclustering microarray gene expression data. Appl Artif Intell 29(4):353–381
9. Ball MG, Qela B, Wesolkowski S (2016) A review of the use of computational intelligence in the design of military surveillance networks. In: Recent advances in computational intelligence in defense and security, Springer, pp 663–693

10. Banerjee S, Hecker JP (2017) A multi-agent system approach to load-balancing and resource allocation for distributed computing. In: First complex systems digital campus world E-conference, Springer, pp 41–54
11. Bayındır L (2016) A review of swarm robotics tasks. Neurocomputing 172:292–321
12. Bianchi L, Dorigo M, Gambardella LM, Gutjahr WJ (2009) A survey on metaheuristics for stochastic combinatorial optimization. Nat Comput 8(2):239–287
13. Blum C, Li X (2008) Swarm intelligence in optimization. Springer, Berlin, pp 43–85
14. Blum C, Roli A (2003) Metaheuristics in combinatorial optimization: overview and conceptual comparison. ACM Comput Surv (CSUR) 35(3):268–308
15. Bonabeau E, Dorigo M, Theraulaz G (1999) Swarm intelligence: from natural to artificial systems. Oxford University Press Inc, New York
16. Botelho S, Alami R (2001) Multi-robot cooperation through the common use of "Mechanisms". In: Proceedings of IEEE/RSJ international conference on intelligent robots and systems (IROS), Maui, USA, pp 375–380
17. Botelho SC, Alami R (1999) M+: A scheme for multi-robot cooperation through negotiated task allocation and achievement. IEEE international conference on robotics and automation, Detroit, USA, vol 2, pp 1234–1239
18. Dias MB, Zlot R, Kalra N, Stentz A (2006) Market-based multirobot coordination: a survey and analysis. Proc IEEE 94(7):1257–1270
19. Dias M (2004) TraderBots: a new paradigm for robust and efficient multirobot coordination in dynamic environments. Ph.D. thesis, Carnegie Mellon University
20. Dias MB, Zinck M, Zlot R, Stentz A (2004) Robust multirobot coordination in dynamic environments. IEEE Proceedings of International Conference on Robotics and Automation, 3435–3442
21. Dorigo M, Birattari M, Stützle T (2006) Ant colony optimization. IEEE Comput Intell Mag 1(4):28–39
22. Dorigo M, Maniezzo V, Colorni A (1996) Ant system: optimization by a colony of cooperating agents. IEEE Trans Syst Man, Cybern Part B (Cybernetics) 26(1):29–41
23. Ducatelle F, Di Caro GA, Pinciroli C, Mondada F, Gambardella L (2011) Communication assisted navigation in robotic swarms: self-organization and cooperation. In: IEEE/RSJ international conference on intelligent robots and systems (IROS), pp 4981–4988
24. Dudek G, Jenkin M, Milios E, Wilkes D (1993) A taxonomy for swarm robots. IEEE/RSJ Int Conf Intell Robots Syst 1:441–447
25. Engelbrecht AP (2005) Fundamentals of computational swarm intelligence. Wiley and Sons Ltd, Chichester
26. Fogel LJ, Owens AJ, Walsh MJ (1966) Artificial intelligence through simulated evolution. Wiley & Sons, New York
27. Von Frisch K (1967) The dance language and orientation of bees. Harvard University Press, Cambridge
28. Garnier S, Gautrais J, Theraulaz G (2007) The biological principles of swarm intelligence. Swarm Intell 1(1):3–31
29. Ge Q, Yang Q, Zhuo P, Liu G, Tang S (2019) Genetic algorithm-based sensor allocation with nonlinear centralized fusion observable degree. IEEE Trans Neural Netw Learn Syst
30. Gerkey BP, Matarić MJ (2002) Sold!: auction methods for multi-robot coordination. IEEE Trans Robot Autom, Spec Issue Multi-Robot Syst 18(5):758–768
31. Gerkey BP, Matarić MJ (2000) MURDOCH: publish/subscribe task allocation for heterogeneous agents. In: Proceedings of the fourth international conference on autonomous agents, Barcelona, Spain, pp 203–204
32. Glover F (1986) Future paths for integer programming and links to artificial intelligence. Comput Oper Res 13(5):533–549
33. Guerrero J, Oliver G (2004) Multi-robot task allocation method for heterogeneous tasks with priorities. In: Proceedings of the 7th international symposium on distributed autonomous robotic systems (DARS), Toulouse, France

34. Gutiérrez A, Campo A, Monasterio-Huelin F, Magdalena L, Dorigo M (2010) Collective decision-making based on social odometry. Neural Comput Appl 19(6):807–823
35. Hefeeda M, Bagheri M (2009) Forest fire modeling and early allocation using wireless sensor networks. Ad Hoc Sensor Wireless Netw 7(3–4):169–224
36. Holland J (1975) Adaptation in natural and artificial systems: an introductory analysis with application to biology. Control and artificial intelligence. MIT Press
37. Hüttenrauch M, Adrian S, Neumann G (2019) Deep reinforcement learning for swarm systems. J Mach Learn Res 20(54):1–31
38. Jevtić A, Gutiérrez A (2011) Distributed bees algorithm parameters optimization for a cost efficient task allocation in swarms of robots. Sensors 11(11):10880–10893
39. Jevtić A, Gutiérrez A, Andina D, Jamshidi M (2012) Distributed bees algorithm for task allocation in swarm of robots. IEEE Syst J 6(2):296–304
40. Jevtić A (2011) Swarm intelligence: novel tools for optimization, feature extraction, and multi-agent system modeling. Ph.D. thesis
41. Jones C, Shell D, Matarić MJ, Gerkey B (2004) Principled approaches to the design of multi-robot systems. In: Proceedings of the workshop on networked robotics, IEEE/RSJ international conference on intelligent robots and systems (IROS 2004)
42. Jones EG, Dias MB, Stentz A (2007) Learning-enhanced market-based task allocation for over-subscribed domains. In: Proceedings of the IEEE/RSJ international conference on intelligent robots and systems. San Diego, CA
43. Kalra N, Stentz A, Ferguson D (2005) Hoplites: a market framework for complex tight coordination in multi-agent teams. In: Proceedings of the international conference on robotics and automation (ICRA), New Orleans, USA, pp 1170–1177
44. Kalra N, Stentz A, Ferguson D (2007) A generalized framework for solving tightly-coupled multirobot planning problems. In: Proceedings of the international conference on robotics and automation (ICRA), Rome, Italy, pp 1050–4729
45. Karaboga D, Akay B (2009) A comparative study of artificial bee colony algorithm. Appl Math Comput 214(1):108–132
46. Khamis A, Hussein A, Elmogy A (2015) Multi-robot task allocation: a review of the state-of-the-art. in cooperative robots and sensor networks, Springer, pp 31–51
47. Kirkpatrick S, Gelatt CD, Vecchi MP (1983) Optimization by simulated annealing. Science 220(4598):671–680
48. Lai K, Rasmusson L, Adar E, Zhang L, Huberman BA (2005) Tycoon: an implementation of a distributed, market-based resource allocation system. Multiagent Grid Syst 1(3):169–182
49. Lee DH (2018) Resource-based task allocation for multi-robot systems. Robot Auton Syst 103:151–161
50. Lee CKH, Ho GTS, Choy KL, Pang GKH (2014) A RFID-based recursive process mining system for quality assurance in the garment industry. Int J Prod Res 52(14):4216–4238
51. Lučić P, Teodorović D (2002) Transportation modeling: an artificial life approach. In: 14th IEEE international conference on tools with artificial intelligence, pp 216–223
52. Macarthur KS, Stranders R, Ramchurn SD, Jennings NR (2011) A distributed anytime algorithm for dynamic task allocation in multi-agent systems. In: Proceedings of the 25th conference on artificial intelligence, pp 701–706
53. Matarić MJ (1994) Interaction and intelligent behavior (No. AI-TR-1495). Massachusetts Institute of Tech. Cambridge Artificial Intelligence Lab
54. Mosteo AR, Montano L, Lagoudakis MG (2008) Multi-robot routing under limited communication range. In: Proceedings of the IEEE international conference on robotics and automation (ICRA), Pasadena, USA, pp 1531–1536
55. Nanjanath M, Gini M (2010) Repeated auctions for robust task execution by a robot team. Robot Auton Syst 58(7):900–909
56. Nanjanath M, Gini M (2006) Dynamic task allocation for robots via auctions. In: Proceedings of the IEEE international conference on robotics and automation (ICRA), Orlando, USA, pp 2781–2786

57. Neapolitan RE, Naimipour K (1996) Foundations of algorithms, D. C. Heath and Company, Lexington, MA, USA
58. Ng KKH, Lee CKM, Chan FT, Lv Y (2018) Review on meta-heuristics approaches for airside operation research. Appl Soft Comput 66:104–133
59. Osman IH, Laporte G (1996) Metaheuristics: a bibliography. Ann Oper Res 63:513–623
60. Ou TC, Su WF, Liu XZ, Huang SJ, Tai TY (2016) A modified bird-mating optimization with hill-climbing for connection decisions of transformers. Energies 9(9):671
61. Pham DT, Ghanbarzadeh A, Koc E, Otri S, Rahim S, Zaidi M (2011) The bees algorithm–a novel tool for complex optimisation. Intelligent production machines and systems-2nd I* PROMS virtual international conference, Elsevier, p 454
62. Poli R, Kennedy J, Blackwell T (2007) Particle swarm optimization. Swarm Intell 1(1):33–57
63. Poole D, Mackworth AK (2010) Artificial intelligence—foundations of computational agents. Cambridge University Press
64. Ramchurn SD, Farinelli A, Macarthur KS, Jennings NR (2010) Decentralized coordination in robocup rescue. Comput J 53(9):1447–1461
65. Ramchurn SD, Polukarov M, Farinelli A, Truong C, Jennings NR (2010a) Coalition formation with spatial and temporal constraints. In: Proceedings of the 9th international conference on autonomous agents and multiagent systems (AAMAS-10), Toronto, Canada, pp 1181–1188
66. Rechenberg I (1973) Evolutions strategie: optimierung technischer Systeme nach Prinzipien der biologischen Evolution. frommann-holzbog, Stuttgart, Germany
67. Robin C, Lacroix S (2015) Multi-robot task allocation and tracking: taxonomy and survey. Auton Robots, pp 1–32
68. Schwarzrock J, Zacarias I, Bazzan AL, de Araujo Fernandes RQ, Moreira LH, de Freitas EP (2018) Solving task allocation problem in multi Unmanned Aerial Vehicles systems using Swarm intelligence. Eng Appl Artif Intell 72:10–20
69. Skiena SS (1998) The algorithm design manual: text. 1. Springer Science & Business Media
70. Smith RG (1980) The contract net protocol: high-level communication and control in a distributed problem solver. IEEE Trans Comput 12:1104–1113
71. Sörensen K, Sevaux M, Glover F (2018) A history of metaheuristics. Handbook of heuristics, pp 1–18
72. Tang Z, Ozguner U (2005) Motion planning for multitask surveillance with mobile sensor agents. IEEE Robot 21:898–908
73. Tang F, Parker LE (2005) Asymtre: automated synthesis of multi-robot task solutions through software reconfiguration. In: Proceedings of the IEEE international conference on robotics and automation, ICRA, pp 1501–1508
74. Tao Y (2018) Swarm intelligence in humans: a perspective of emergent evolution. Physica A 502:436–446
75. Teodorović D, Dell'Orco M (2005) Bee colony optimization—a cooperative learning approach to complex transportation problems. Advanced OR and AI methods in transportation. In: Proceedings of 10th meeting of the EURO working group on transportation, p 5160
76. Tkach I, Jevtić A, Nof SY, Edan Y (2013) Automatic multi-sensor task allocation using modified distributed bees algorithm. In: IEEE international conference on systems, man, and cybernetics (SMC), Manchester, England, pp 1401–1406
77. Tkach I, Jevtić A, Nof S, Edan Y (2018) A modified distributed bees algorithm for multi-sensor task allocation. Sensors 18(3):759
78. Walsh WE, Wellman MP (1998) A market protocol for decentralized task allocation. In: Proceedings of the international conference on multi-agent systems, pp 325–332
79. Wedde HF, Farooq M, Zhang Y (2004) Beehive: an efficient fault-tolerant routing algorithm inspired by honey bee behavior. Ant Colony, optimization and swarm intelligence, Lecture notes in computer science, vol 3172. Springer-Verlag, pp 83–94
80. Werger B, Matarić MJ (2000) Broadcast of local eligibility for multi-task observation. In: Parker LE, Bekey G, Barhen J (eds) Distributed autonomous robotic system, Springer, Berlin, Germany, vol 4, pp 347–356

81. Wolski R, Plank JS, Brevik J, Bryan T (2001) Analyzing market-based resource allocation strategies for the computational grid. Int J High Perform Comput Appl 15(3):258–281
82. Yang XS (2005) Engineering optimizations via nature-inspired virtual bee algorithms. Artif Intell Knowl Eng Appl Bioinsp Appr 3562:317–323
83. Yedidsion H, Zivan R, Farinelli A (2018) Applying max-sum to teams of mobile sensing agents. Eng Appl Artif Intell 71:87–99
84. Zhang J, Li W, Han N, Kan J (2008) Forest fire allocation system based on a ZigBee wireless sensor network. Front For China 3(3):369–374
85. Zlot R, Stentz A (2006) Market-based multirobot coordination for complex tasks. Int J Robot Res. Special issue on the 4th international conference on field and service robotics, 25(1):73–101

Chapter 6
Extended Examples of Single-Layer Multi-sensor Systems

This chapter describes three examples of multi-agent systems: a multi-sensory security system for supply networks, a multi-agent approach to solve the travelling salesman problem and multiple police officers allocated to crime incidents in law enforcement problem. These examples are evaluated by task allocation algorithms to demonstrate the methods described in Chap. 5. The comparison of the performances of nine state-of-the-art algorithms for these case studies is conducted in terms of tasks completion times and the number of unallocated tasks. Scalability and the influence of bias parameters of HDBA were analyzed for different numbers of sensors and tasks.

6.1 Security of Supply Networks

Supply networks are systems responsible for moving products or services and represent a complex network of interrelated entities, including suppliers, manufacturers, retailers, and customers [34]. As supply networks have become more global, they become more vulnerable to security risks [33, 30, 48]. The security spans over the entire supply network and includes facility security, information security, and cargo security among others [33]. Moreover, with the growth of security threats and cargo theft in the world, monitoring and protecting the supply network became a priority concern in most countries [9] and especially in South America [10, 50].

While security has been the primary focus of several organizations involved in supply network operations (firms, governments, carriers, and consumers), it has been traditionally treated by policies and regulations [9, 33, 40, 45, 55]. Firms and businesses have primarily focused on theft issues and asset security; governments have focused on restricting the flow of illegal items and preventing terrorism [9, 33, 45]. To effectively meet the demands of a secure supply network in today's environment, a more comprehensive, technological, and integrated security focus is required for risks allocation and prevention, extending beyond procedural and organizational approaches [9, 33].

© Springer Nature Switzerland AG 2020
I. Tkach and Y. Edan, *Distributed Heterogeneous Multi Sensor Task Allocation Systems*,
Automation, Collaboration, & E-Services 7,
https://doi.org/10.1007/978-3-030-34735-2_6

The common technological approach for dealing with supply chain security and management includes radio frequency identification (RFID) technology, which provides a common way to obtain information on individual items (e.g., cars and trucks, containers, pallets; [39, 41, 58]; Sidorov et al. [46]) and is typically used for supply chain management [12] and manufacturing [7, 26, 36]. RFID is an auto-ID technology that harnesses electromagnetic fields for using radio waves to assist in data exchange between tags and readers. Other auto-ID technologies include barcodes, biometrics, and smart cards [2].

RFID has been applied in different logistics areas improving shipping, distribution, and manufacturing processes and in industries such as automotive, military, retailing, agriculture, healthcare, pharmaceutical, and security [3, 4, 37]. In supply chain management, RFID tags are used to track products throughout the supply network—from supplier delivery to warehouse stock, and point of sale ([22]; Fig. 6.1). A central database records product movement, which manufacturers or retailers can later query for location, delivery confirmation, or theft prevention [54]. RFID can improve supply chain management efficiency and ease of use. However, the RFID-based supply network is exposed to security and privacy challenges. An organization that implements RFID in its supply network does not want competitors to track its shipments and inventory [54]. Therefore, a secure RFID system must avoid eavesdropping, traffic analysis, spoofing, and denial of service [15].

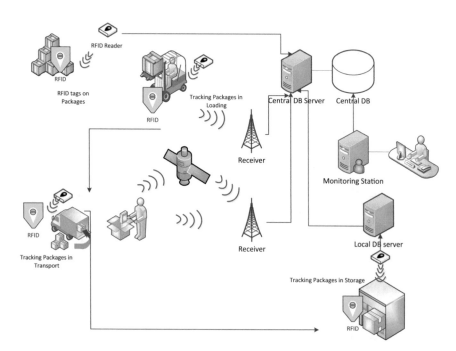

Fig. 6.1 RFID-based supply network management system scheme (after Tkach et al. [50])

The existing gap in the technological approach for dealing with supply network security is that there is no methodology or framework established for the design of a system to detect threats with dynamic and unpredictable nature.

As RFIDs have known limitations such as limited communication bandwidth, reliability issues, tag memory, range coverage, and preprogrammed information usage [13, 49], monitoring the supply network security via distributed sensors and sensor networks additional to RFIDs can contribute to improved performance [14] and cope with unpredictable security tasks. Combining additional multiple and different types of sensors (e.g., visible, infrared, and night vision among others) can further increase performance and reliability [5]. Moreover, collaboration among different sensors through information sharing can significantly enhance the overall performance of supply networks [27]. Computer-supported assignment, allocation, and priority logic and negotiation procedures can automatically resolve collaboration conflicts arising from competition for a limited number of sensors [56].

The major limitation is that there is no technological approach for real-time threat monitoring in supply networks. This book shows how to use a distributed heterogeneous multi-sensor task allocation system to enable supply network security (Fig. 6.2).

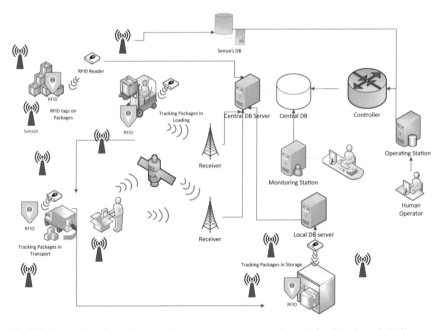

Fig. 6.2 Sensor based supply network management system scheme (after Tkach et al. [50])

6.1.1   The Use of Algorithms to Enable Supply Network Security

Three types of sensors are considered with different allocation ranges and performances. Experiments were performed for a square-shaped simulation arena, but it can be easily extended to relatively complicated geometries. In addition, three deployment strategies have been considered based on Jeong and Nof [23] as shown in Fig. 6.3: (a) deterministic deployment, i.e., grid deployment; (b) random deployment, i.e., uniformly distributed random deployment; and (c) random and biased deployment, i.e., normally distributed random deployment. Deployed sensors are marked as blue circles and tasks are marked by dots (black dots represent tasks that were completed, and red dots represent non-completed tasks). In these figures, there are 200 tasks (final distribution of tasks is described on the map) and 100 sensors distributed over the map. The effects of various sensor densities were examined. The choice of algorithms was made to include different approaches; specifically, we

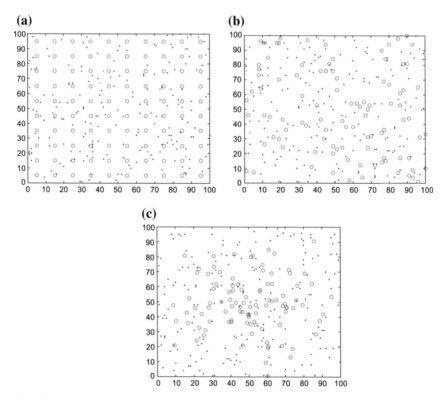

Fig. 6.3 Simulation map of the arena with three sensors distributions—**a** grid distribution, **b** uniformly distributed random deployment, **c** normally distributed random deployment; deployed sensors are marked as blue circles, incomplete tasks are marked by red dots, and completed tasks are marked by black dots (after Tkach et al. [52])

Table 6.1 Summary of parameter values for simulation analysis

Parameter	Values
Area dimensions	100×100 m
Number of sensors ('bees')	20, 40, 60, 80, 100
Number of tasks	200
Task completion time	Randomly distributed from 0 to 10 min
Simulation duration	200 steps
Tasks location	Uniformly distributed at random
Control parameters	$\alpha = \beta = \gamma = 1, \alpha = 2\beta = 2\gamma, \alpha = 2\beta = \gamma, \alpha = 2\beta = 0.5\gamma$
Sensor location	Uniformly distributed random
	Normally distributed random
	Grid deployment
Task arrival time	Every 1 step
Sensor range coverage	Predefined from 15 to 45 m

compared greedy versus heuristic and swarm-based versus market-based approaches. Accordingly, the algorithms compared were DBA, market-based algorithm, greedy algorithm, and Bee System. Due to the complexity of the problem a comparison was made with decentralized versions of market-based and greedy algorithms. The centralized version of these algorithms was implemented before in Tkach et al. [51] but yielded poor results.

The parameters that were used for the simulation are presented in Table 6.1. Tasks were uniformly distributed in the arena based on Rowaihy et al. [44] and Byers and Nasser [8]. The sensors' range values were predefined to cover 15–45 m relative to the size of the arena (100×100 m). An inherent noise in each sensor was introduced according to typical operational distance ranges of known commercial sensors [59] relative to the arena size: acoustic sensor, seismic sensor, and forward looking infrared radar (FLIR). The allocation range of these sensors is given by:

$$R_{seismic} = 0.8 \times R_{acoustic} = 1.25 \times R_{FLIR} \tag{6.1}$$

The numerical computations were performed on a PC with 2.90 GHz CPU, and 12 GB of RAM, using Matlab R2015a software.

6.1.2 Performance Measures

The following performance measures are analyzed to compare the different allocation algorithms:

1. System performance as defined in (5.1).
2. Tasks completion time as defined in (5.9).

3. Number of unallocated tasks is defined by:

$$\varepsilon = \sum_{i=1}^{M} task_i; \quad \begin{array}{l} task_i = 1 \text{ if } \sum_{k=1}^{N} V_{ik} = 0 \\ else \ task_i = 0 \end{array} \tag{6.2}$$

where ε is the number of unallocated tasks, M is the total number of tasks in the system, N is the total number of sensors in the system, i is the index of the current task, and V_{ik} is the kth sensor's performance on the ith task.

4. Number of tasks allocated to a sensor k is defined by:

$$\chi_k = \sum_{i=1}^{M} task_i; \quad \begin{array}{l} task_i = 1 \text{ if } V_{ik} > 0 \\ else \ task_i = 0 \end{array} \tag{6.3}$$

where χ_k is the number of treated tasks by the kth sensor, M is the total number of tasks in the system, i is the index of the current task, and V_{ik} is the kth sensor's performance on the ith task.

The mean values that were obtained from 100 independent runs of HDBA, DBA, BS, market-based, and greedy algorithms, were compared at the statistical confidence level of 95%.

6.1.3 Evaluation Scenarios

The algorithm's performance is evaluated for the scenarios of:

1. Task allocation (Sect. 6.1).
2. Traveling salesman problem (Sect. 6.2).
3. Law enforcement problem (Sect. 6.3).

6.1.4 Results and Discussion

Results revealed, as expected, that as the number of sensors increased, the mean number of unallocated tasks decreased for all of the algorithms (from 72.53 to 4.74, from 78.25 to 5.88, from 81.14 to 6.07, from 83.13 to 5.98, and from 87.65 to 8.61 in HDBA, DBA, market-based, BS, and greedy algorithms, respectively; Fig. 6.4). By adding more sensors, the impact of their relative contribution decreases due to the overlapping range coverage (e.g., the difference in the mean number of unallocated

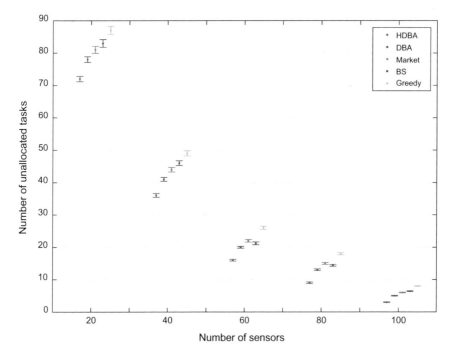

Fig. 6.4 Number of unallocated tasks for a different number of deployed sensors for each algorithm with uniformly distributed random sensors deployment. The dots represent mean values of 100 independent runs and the bars represent 95% confidence intervals

tasks between 20 sensors and 40 sensors and between 80 and 100 sensors is 36 and 6, respectively). Therefore, optimal sensor density can be changed with the desired geometry of the arena and must be obtained through simulations. The bio-inspired algorithms HDBA and DBA resulted in a lower number of unallocated tasks than the market-based and greedy algorithms by 45.9% on average (from 29.31, CI = 0.41 to 42.76, CI = 0.56, $p < 0.05$) and with maximum of 81.7% difference (from 4.74, CI = 0.097 to 8.61, CI = 0.11, $p < 0.05$). The high number of unallocated tasks (72) when using 20 sensors was due to the limited coverage range of the sensors.

The best performance of HDBA was obtained for sensors with a uniform distribution (Fig. 6.5), resulting in the shortest task completion time with 17.1 and 26.3% improvement from the grid and normal distributions, respectively ($p < 0.05$). The normally distributed random sensor deployment achieved lower performance due to the fact that most of the sensors were deployed in the center of the arena. Thus, tasks that were generated in the edges of the arena were treated with lower performance. The performance changed as time advanced since sensors were allocated to the high priority tasks that were generated in a particular simulation cycle.

The task completion times of 100 sensors using HDBA, DBA, market-based, BS, and greedy algorithms are shown in Fig. 6.6. The HDBA algorithm resulted in the lowest task completion time (291.21 min, CI = 3.42, which is 6.6% lower than the

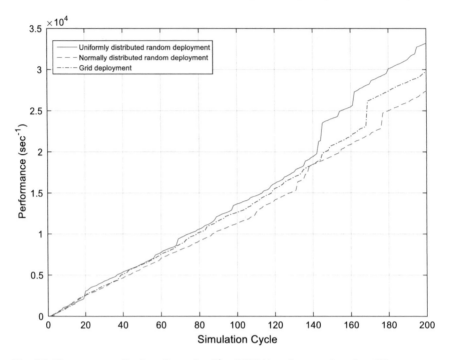

Fig. 6.5 Heterogeneous distributed bees algorithm (HDBA) performance based on different sensor distributions (after Tkach et al. [52])

second–best algorithm, 310.12 min, CI $= 3.47$, and 18.3% when compared to the greedy algorithm, 344.41 min, CI $= 3.52$, all with $p < 0.05$).

The greedy algorithm resulted in the longest completion time, due to its intrinsic greedy logic, which resulted in the constant occupation of the best performing sensors without attempts to create allocation combinations that optimize the overall performance. Also, it must be noted that despite the shorter completion time (291.21 min), HDBA resulted in a higher average number of tasks (43.22, CI $= 0.34$) that were treated by each sensor when compared to the market-based algorithm (38.62, CI $= 0.32$, with 336.71 min run time) and greedy algorithm (31.17, CI $= 0.31$, with 344.41 min run time). In this sense, HDBA has a slight disadvantage in excessive sensor occupation (Fig. 6.7), which can result in a higher failure rate and eventually reduce the overall system availability; however, these risks could be reduced by a more redundant sensor design. For example, if the system is designed with several redundant sensors in addition to the operational sensors, then the system's availability will increase.

The simulation results are summarized in Table 6.2.

Collaboration was preferred for task completion. Since sensors collaborate to complete tasks, each task was performed by many sensors. As a result, each sensor treated a high number of tasks through collaboration.

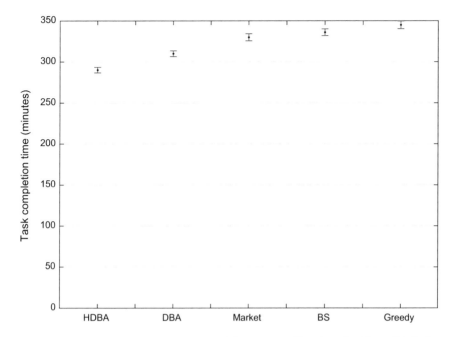

Fig. 6.6 Average performance comparison of four sensor allocation algorithms. Black dots represent mean values of 100 independent runs and red bars represent 95% confidence intervals

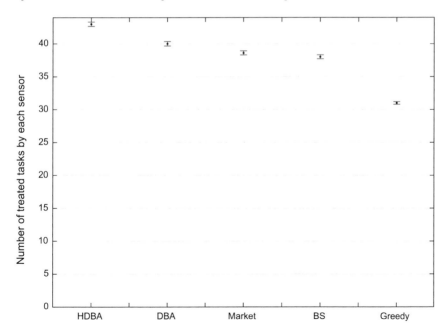

Fig. 6.7 The average number of treated tasks by each sensor in four algorithms. Black dots represent mean values of 100 independent runs and red bars represent 95% confidence intervals

Table 6.2 Summary of results for different algorithms evaluation with 100 independent runs

Measured value	HDBA	Standard deviation	DBA	Standard deviation	BS	Standard deviation	Market	Standard deviation	Greedy	Standard deviation
Number of treated tasks by each sensor	43.22[*]	3.37	39.99[*]	3.42	37.48[*]	4.12	38.62[*]	3.25	31.17[*]	3.12
Number of unallocated tasks	29.31[*]	4.07	31.85[*]	3.96	33.79[*]	5.16	34.76[*]	4.23	42.76[*]	8.99
Task completion time	291.21[*]	34.23	310.12[*]	34.79	339.22[*]	41.88	336.71[*]	31.30	344.41[*]	35.21

[*]Statistically significant results

6.1.5 Scalability Evaluation of HDBA

Scalability was defined by Rana and Stout [42] and Ito et al. [21] as an ability to cope with the changes in the number of agents. Following this definition, in order to evaluate the system's scalability, simulation analysis was conducted for different numbers of both sensors and tasks. In the initial setup, 80 sensors were uniformly distributed in the arena. Twenty additional sensors were introduced into the system in two phases. In the last phase, 10 sensors were removed. The number of tasks has increased throughout the simulation from 0 to 200.

The system was able to scale-up with different numbers of sensors and tasks, and to maintain the ability to allocate sensors to the upcoming tasks (Fig. 6.8). In all three phases that were simulated by the addition and subtraction of sensors during the simulation, the system continued to allocate sensors and achieved an average value of tasks completion time of 312.55 min, CI = 3.53, which is 7.3% higher than a system initially with 100 distributed sensors [52].

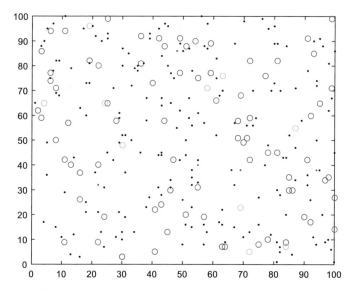

Fig. 6.8 Simulation map of the arena with subsequently introduced sensors—initially deployed sensors are marked as blue circles, deployed sensors at phase 1 are marked as green circles, deployed sensors at phase 2 are marked as yellow circles, removed sensors at phase 3 are marked by red circles, incomplete tasks are marked by red dots, and completed tasks are marked by black dots (after Tkach et al. [52])

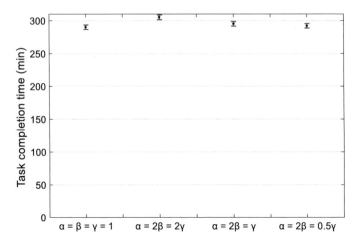

Fig. 6.9 Average performance comparison of HDBA algorithm with different α, β, and γ values. $\alpha = \beta = \gamma = 1$; $\alpha = 2\beta = 2\gamma$; $\alpha = 2\beta = \gamma$; $\alpha = 2\beta = 0.5\gamma$. Black dots represent mean values of 100 independent runs and red bars represent 95% confidence intervals

6.1.6　Influence of Bias Parameters on HDBA Behavior

The control parameters, α, β, and γ, provide a mechanism to adjust the sensor swarm behavior. The values of α, β, and γ were changed in order to bias the resulting sensors' allocation and to test the system's performance for different values of these parameters: $\alpha = \beta = \gamma = 1$, $\alpha = 2\beta = 2\gamma$, $\alpha = 2\beta = \gamma$, $\alpha = 2\beta = 0.5\gamma$.

The range for these parameters was set according to Jevtić and Gutiérrez [24] and Dorigo et al. [11]. The algorithm was able to converge in all of the scenarios.

A simulation with different values of α, β, and γ was conducted and revealed that the setup of $\alpha = \beta = \gamma = 1$ achieved the best results (291.21 min) in terms of task completion time (Fig. 6.9), outperforming systems with $\alpha = 2\beta = 2\gamma$, $\alpha = 2\beta = \gamma$ and $\alpha = 2\beta = 0.5\gamma$ by 5.2, 1.8, and 0.7%, respectively ($p < 0.05$). These parameters play an important role in defining the system's overall behavior. For example, they can be set to prioritize closer tasks and reduce traveled distances at the expense of the execution time (a useful feature for mobile multi-sensor systems). Allocating the tasks according to their priorities would decrease the execution time of crucial tasks at the expense of longer execution of less important tasks [52].

6.2　The Traveling Salesman Problem

The Traveling Salesman Problem (TSP) is an NP-hard problem that is defined as finding the shortest possible route between cities in a closed network, visiting each city exactly once, and returning to the origin city [20, 53]. The TSP is also defined as the problem of finding the shortest route to visit a set of customers by a single salesman and to return. It is the most known distributed constraint optimization problem, and due to its simple representation, it is used as a benchmark for numerous optimization methods [17, 18] (Fig. 6.10).

Following previous chapters, we test HDBA for two TSP benchmark instances Berlin52 and A280 using TSPLIB as a reference. TSPLIB is an online library developed by the University of Heidelberg in Germany that contains several samples of TSP and similar related problems ranged on a list of different files. It has become a standard reference in modern research and the documentation can be obtained in Reinelt [43], found in TSPLIB: http://www.iwr.uni-heidelberg.de/iwr/comopt/soft/TSPLIB95/TSPLIB.html.

The coordinates of cities are decimal numbers including negative values and the distances between the cities are computed according to the Pythagorean Theorem.

In this evaluation, TSP is symmetric, with the distance between two neighboring cities a and b equal to the distance between b and a. The optimal solution can be derived by calculating all possible paths and then choosing the shortest one. This requires an unlimited amount of calculation power for a large number of cities. Due to the symmetrical nature of the problem, the total number of possible tours in TSP is given by the expression $(n - 1)!/2$ (where the tour will consist of n cities).

If we assume that we can calculate one tour per nanosecond (10^{-9} s) [47], the time consumption will be as seen in Table 6.3. We can see that for 20 cities the exact computation of the solution is almost impractical and the time for calculation of all possible tours will be more than a year [20].

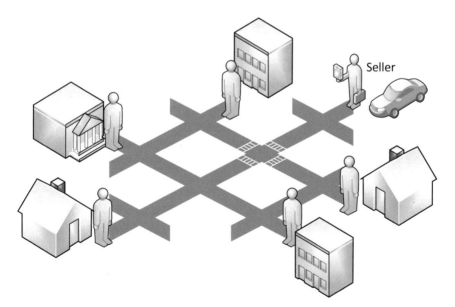

Fig. 6.10 The traveling salesman problem illustration. The goal is to visit each city once with the shortest possible route

Table 6.3 The number of possible tours and processing time in years for different numbers of cities

n	Number of possible tours [(n − 1)!/2]	Time in years
3	1	≈ 3.17058E−17
4	3	≈ 9.51173E−17
5	12	≈ 3.80469E−16
6	60	≈ 1.90235E−15
7	360	≈ 1.14141E−14
8	2520	≈ 7.98985E−14
9	20,160	≈ 6.39188E−13
10	181,440	≈ 5.75269E−12
15	43,589,145,600	≈ 1.38203E−06
20	60,822,550,204,416,000	≈ 1.928425815
25	310,224,200,866,620,000,000,000	≈ 9,835,897.301
30	4,420,880,996,869,850,000,000,000,000,000	≈ 1.40167E + 14
35	147,616,399,519,802,000,000,000,000,000,000,000,000	≈ 4.68029E + 21
40	10,198,941,040,598,700,000,000,000,000,000,000,000,000,000,000	≈ 3.23365E + 29
45	1,329,135,787,394,220,000,000,000,000,000,000,000,000,000,000,000,000,000	≈ 4.21413E + 37
50	304,140,932,017,134,000,000,000,000,000,000,000,000,000,000,000,000,000,000,000,000,000	≈ 9.64302E + 45

Two evaluations were conducted, the first instance is Berlin52 with 52 locations in Berlin and the second and largest instance has 280 cities [43]. Due to its complexity, this instance requires more computational time. The compared algorithms yield good

Table 6.4 Summary of the results for each algorithm

	Optimal	HDBA	DBA	BS	ACO	GA	SA	Greedy
Berlin 52	7542	7577	7885	7634	7575	7542	9269	8356
A280	2579	2936	2950	2784	3092	3218	5743	4109

results in handling this NP-hard optimization problem. Table 6.4 shows the computational results. HDBA provides good results for the two instances; it shows adequate performance over the small instance of Berlin52 and robustness over the large A280. HDBA managed to solve TSP without any prior knowledge of the solution space or the help of any local search routine. The algorithm is robust and maintains a steady improvement in behavior, it always gives solutions and solutions get better over time. HDBA resulted in a second-best algorithm for the large instance of A280 after BS. It is prone to falling into a local optima due to its probabilistically which updates the solution according to the current best path.

In HDBA every partial traveling salesman tour has two main attributes: (a) the total length, and (b) the number of 'bees' that are using the partial route. Both normalized values of these attributes are defined in the following way: (a) both normalized values can take any value between 0 and 1; (b) the smaller the total length normalized value, the better the partial tour; (c) the bigger the number of 'bees' normalized value, the better the partial tour. After each iteration of complete tour generation, the best 'bee' is selected and its tour is given a better performance. The probability that the partial route will be chosen by any agent deciding to choose the new route is based on Eq. (5.13):

$$
p_{ik} = \frac{F_i^\alpha \left(\frac{1}{D_{ik}}\right)^\beta V_{ik}^\chi}{\sum_{j=1}^{M} F_j^\alpha \left(\frac{1}{D_{jk}}\right)^\beta V_{jk}^\chi}
$$

where F_i is the number of agents selecting the tour i; D_{ik} is the distance from city i to city k; V_{ik} is the updated performance of the tour from city i to city k.

Pseudocode of the HDBA algorithm for TSP

1. Initialize:
   Set $F=1$
   Set $V=1$
   Randomly place N 'bees' on cities (Figure 6.11).
2. Until *termination condition not met*
   For $i=1$ to M do
      For $k=1$ to N do
         Calculate distances of agents from nearest cities: D_i
         Calculate probabilities for each sensor p_{ik}
         Apply wheel-selection rule
         Move 'bees' to the selected cities
         Calculate the number of agents selecting the tour i: F_i
         Select the best 'bee' and update its tour performances: V_{ik}
3. Finalize:
   Select the shortest tour
   Stop

See Fig. 6.11.

Figures 6.12, 6.13, 6.14, 6.15, 6.16 and 6.17 illustrate the convergence of HDBA to the solution in Berlin52 and A280 instances. Figure 6.12 represents the initial state of the Berlin52 instance, where blue dots represent the cities and the scales on the axes represent the distances. Figure 6.13 represents the intermediate state of the Berlin52 simulation process using HDBA. Lines of different colors represent different tours

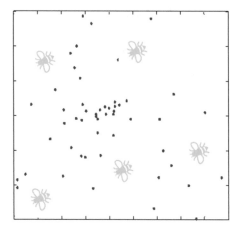

Fig. 6.11 'Bees' placement in random cities in the initial state of TSP

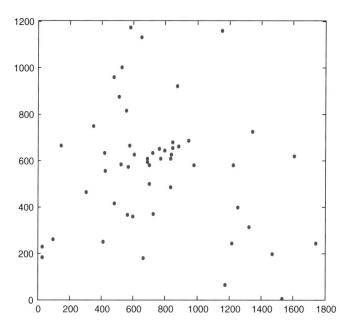

Fig. 6.12 Initial state of Berlin52 simulation

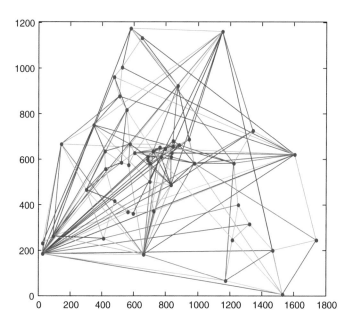

Fig. 6.13 Intermediate state of Berlin52 simulation. HDBA generated many possible paths

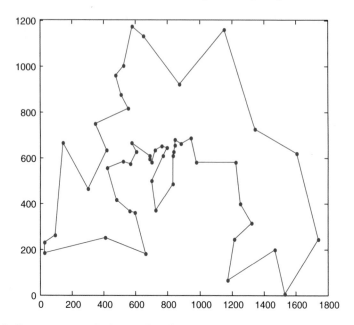

Fig. 6.14 Convergence to a single path of Berlin52

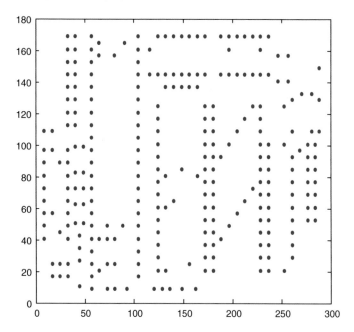

Fig. 6.15 Initial state of A280 simulation

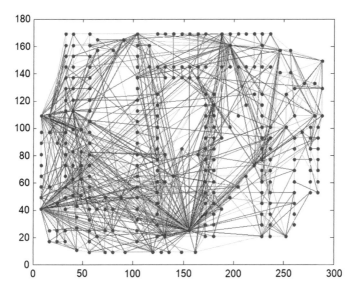

Fig. 6.16   Intermediate state of A280 simulation. HDBA generated many possible paths

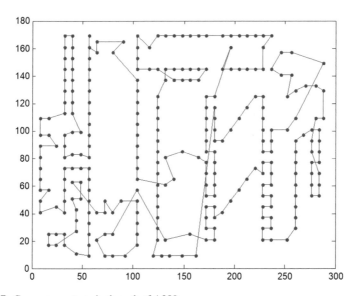

Fig. 6.17   Convergence to a single path of A280

generated by HDBA. Figure 6.14 represents the final state of the simulation. In this state, HDBA converged to the final solution that was the best during the simulation.

Figure 6.15 represents the initial state of A280 instance, where blue dots represent the cities and the scales in the axes represent the distances. Figure 6.16 represents the intermediate state of Berlin52 simulation process using HDBA. Lines of different colors represent different tours generated by HDBA. Figure 6.17 represents the final state of the simulation. In this state, HDBA converged to the final solution that was the best during the simulation.

Table 6.4 summarizes the results of the six algorithms for the two simulated instances for 100 simulation cycles. Heuristics that serve as benchmarks for TSP (i.e., GA and ACO) were chosen. The results of BS, ACO, GA, and Greedy algorithms were obtained from Haroun and Jamal [19], Lučić and Teodorović [32, 31], and Kongkaew and Pichitlamken [28]. In the Berlin52 instance, GA achieved the best result of 7542. HDBA was able to solve this instance with the result of 7577. In the A280 instance, HDBA achieved the second-best result of 2936, indicating its fitness to solve a large and complex TSP instance.

6.3   The Law Enforcement Problem

The law enforcement problem (LEP) is an NP-hard assignment problem where a team of cooperative heterogeneous agents, each with a possibly different set of skills, has a common goal [16, 29, 57]. In LEP police officers with heterogeneous skills conduct routine patrols and perform tasks in response to reported incidents. The law enforcement tasks associated with LEP require specific combinations of agent skills. LEP was extensively analyzed by Nelke [38], who employed and compared two algorithms FMC_TA[H+] and SA, including optimization of the values of prices and penalties. Therefore, in the current book we used these optimized algorithms for comparison with HDBA.

The goal of this chapter is to meet the challenge of solving the realistic LEP problem based on real police logs by applying and comparing HDBA to FMC_TA[H+] and SA, which were found by Amador et al. [1] to be most successful in solving LEP.

Greedy algorithm was analyzed both analytically and numerically by Nelke [38]. These analyses revealed the disadvantage of the greedy algorithm, which allocated events to the closest agents that were on patrol when they appeared, thus resulting in worse allocation.

In such realistic applications as LEP, avoiding dependence on a central dispatch for coordinating task allocation is preferred. In a disaster scenario, for instance, communication lines to a central location may break down and a single-point-of-failure is preferably avoided, especially in malicious scenarios (e.g., a terror attack).

The importance of each task varies from low (e.g., noise complaint) to high (e.g., murder). The workload associated with each task, indicating the amount of work to be completed for the incident to be processed, may vary as well. Multiple officers with heterogeneous skills may work together on important tasks to share the workload and improve response time. To deal with LEP, one must allocate police officers to dynamic tasks whose locations, arrival times, and importance levels are unknown a priori. The three methods were compared in this section for five different performance measures that are commonly used by law enforcement authorities.

Tasks may differ in complexity, importance, and urgency. In some cases, the task is too large or complex to be handled by a single agent, while in other cases the task may require multiple skills or fields of expertise that no single agent possesses. Furthermore, by combining the capabilities and joint effort of multiple agents, the time required to complete a task may be significantly reduced, even if the task does not strictly require such collaboration. An example of LEP with four agents allocated to two tasks is illustrated in Fig. 6.18.

Consider the example presented above that includes an artificial city patrolled by four police units (agents), a_1, a_2, a_3 and a_4, respectively.

Agents derive utility when they perform a task in a time unit. For a task Ψ_i that arrives at time t_{Ψ_i}, we denote by $w(\Psi_i)$ the workload for task Ψ_i in time units (i.e., the number of units required to handle the task). In a real-world domain, it symbolizes the amount of time it takes a traffic officer or a detective to write down a report or to conduct an investigation of a crime scene respectively. The utility for handling the task is $U(\Psi_i)$. The utility derived from patrolling is $U(p)$. The time that it takes agent a_k to reach task Ψ_i is denoted by $\rho(a_i, \Psi_i)$. Obviously, an agent will decide to handle a task only if $U_{i,t}(\Psi_i) > U_{i,t}(p)$.

For a single task Ψ_i, if $U_{k,t}(\Psi_i) \leq U_{k,t}(p)$ when for each $k \in \{1, 2, 3, 4\}$, then the optimal allocation for the team is to avoid handling the events and keep patrolling. For any other case, where there is a single task Ψ_i, and for at least one of the agents $U_{k,t}(\Psi_i) \geq U_{k,t}(p)$ then $U = \Sigma U_k$.

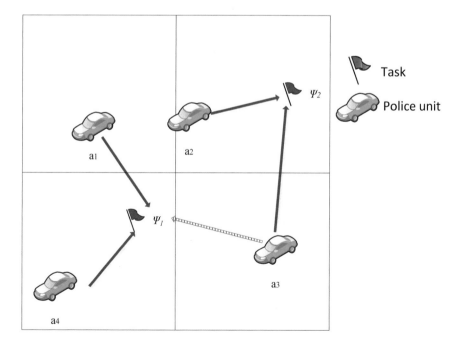

Fig. 6.18 LEP example with 4 police units and 2 tasks

Following example illustrated in Fig. 6.18, upon the appearance of the two tasks Ψ_1 and Ψ_2, agents a_1 and a_4 will be allocated to task Ψ_1, agent a_2 and agent a_3 will be allocated to task Ψ_2. The total utility derived in this case will be $U = U_{1,t}(\Psi_1)$ $+ U_{4,t}(\Psi_1) + U_{2,t}(\Psi_2) + U_{3,t}(\Psi_2)$. This example, while very simple, illustrates a possible allocation of agents to tasks and the total utility derived.

The utility an agent derives from performing a task is affected by many factors, including the severity and urgency of the task, travel distance to the task's location, current agent occupation, consequences of abandoning the current task, and agent skills.

In solving LEP, agents to be allocated have together a set of j unique skills S, with each agent k having a subset $S_k \in S$ of skills.

The combined effort required for completing each LEP task Ψ_i is a workload set W_i, where $w^s(\Psi_i) \in W_i$ ($w^s(\Psi_i) \in R$) specifies the workload of a specific skill s that should be applied to Ψ_i. Thus, Ψ_i is completed only if all the work specified by members in W_i has been performed by agents who possess the required skills. Formally, the total workload for completing a task Ψ_i is:

$$W_i = \sum_s w^s(\Psi_i) \tag{6.4}$$

The allocation specifies not only what tasks an agent has to perform but also what skills they have to use in the process. If the agent is required to use more than one skill in performing the task, they must do so sequentially. If multiple agents are performing different skills, they can do so concurrently. Formally, allocation of tasks to agents in LEP is denoted by an $n \times m \times k$ matrix X where entry x_{iks} is the fraction of task Ψ_i assigned to agent k, utilizing the relevant skills. The schedule of each agent must include, besides the task being performed as well as the start and end times, the skill the agent is utilizing. Thus, each member of the schedule σ^k of agent k includes (Ψ_i, s, t, t'), specifying, respectively, the task, the skill being utilized, and the start and end times for applying skill s on task Ψ_i by agent k. In addition, there are inter-skill constraints that require concurrence between police officers with different skills. The performance of task Ψ_i depends on the number of agents that work simultaneously on that task. Thus, capability function is defined for a vector $\vec{q} \in N$ that specifies the number of agents with each skill working concurrently on a task, i.e., the l'th entry in $\vec{q}$ represents the number of agents with skill s_l working on the task concurrently. Each agent can be counted only once, i.e., they cannot utilize multiple skills simultaneously. The result of the function $Cap(\Psi_i, \vec{q})$ is a vector $\vec{g}$ specifying for each skill the utility derived by an agent performing it, taking under consideration the number of agents performing this skill and the inter-skill constraints. Let $d(\Psi_i, \vec{q})$ be the time duration that $\vec{q}$ represents the set of agents working simultaneously on task Ψ_i. Thus, $\frac{d(\Psi_i, \vec{q})}{w(\Psi_i)}$ is the relative portion of time that the set of agents specified by $\vec{q}$ are working on Ψ_i. Denote by $\vec{Q}$ the set of all possible vectors $\vec{q}$. The utility derived by the agents for completing Ψ_i is:

$$U' = \sum_{\vec{q} \in Q} \frac{d(\Psi_i, \vec{q})}{w(\Psi_i)} \sum_{l=1}^{i} q[l]g[l] \tag{6.5}$$

where $q[l]$ and $g[l]$ are the l'th entry in vectors $\vec{q}$ and $\vec{g}$, respectively, a workload $w(\Psi_i)$ indicates how much work (in time units) must be performed to complete the task Ψ_i.

The utility derived for completing task Ψ_i starting at time t_{ψ_i} depends on the capability of the agents performing the task and the soft deadline function $\Delta(\Psi, t)$ $= \beta^{\gamma(t_{\psi_i} - t_{i_{arrival}})}$. The discounted utility for performing task Ψ_i with arrival time at time $t_{i_{arrival}}$ and which is initially handled at time t_i is:

$$U'(\Psi_i) = \beta^{\gamma(t_{\psi_i} - t_{i_{arrival}})} \sum_{\vec{q} \in Q} \frac{d(\Psi_i, \vec{q})}{w(\Psi_i)} \sum_{l=1}^{i} q[l]g[l] \tag{6.6}$$

where $\beta \in (0, 1)$ and $\gamma \geq 0$ are constants.

When a new task arrives, the current task (if any) being performed by agent k is denoted CT_k and the current skill that is used by agent k for CT_k is denoted CS_k. Agents can interrupt the performance of their current task. The penalty for task interruption is:

$$\pi\left(\Psi_i, \Delta w_i^{CS_k}\right), \tag{6.7}$$

which depends on the task Ψ_i and the amount of work for skill CS_k completed when the task is interrupted. The adjusted penalty for task Ψ_i decreases exponentially with to a minimum value:

$$\pi\left(\Psi_i, \Delta w_i^{CS_k}\right) = \begin{cases} max\left\{I(\Psi_i)c^{w_i^{CS_k} - \Delta w_i^{CS_k}}, \Xi I(\Psi_i)\right\} & if \ w_i^{CS_k} - \Delta w_i^{CS_k} > 0 \ and \ CS_k \in S_k \\ 0 & otherwise \end{cases} \tag{6.8}$$

where $c \in (0, 1)$ and $\Xi > 0$ are constants and $\Xi I(v_i)$ is the minimum penalty.

The total utility derived for performing Ψ_j is thus:

$$U(\Psi_i) = U'(\Psi_i) - \sum_{a_i : \Psi_i^i \neq CT_i} \pi(CT_k, \Delta w_i^{CS_k}) \tag{6.9}$$

Table 6.5 illustrates the different LEP elements as a task allocation problem, market clearing problem, and swarm intelligence problem.

The experimental design resembles a realistic LEP in the city of Beer-Sheva in Israel. To be able to compare HDBA with the best performing algorithms of this benchmark and to make a fair comparative analysis, two simulation settings were

Table 6.5 Modeling law enforcement as task allocation and market clearing problems

LEP	Task allocation	Market clearing	Swarm intelligence
Police units	Agents	Buyers	Agents 'bees'
Missions	Tasks	Goods	Tasks
Utility (of a police unit performing a task)	Utility (of assigning an agent to a task)	Preference (value of a goods to a buyer)	Probability (of allocating an agent to a task)

conducted. In the first set, the values of the swarm size and the maximum number of cycles of HDBA were set according to the values analyzed in LEP by Nelke [38], where the highest number of agents was 25. In the second set, the swarm size was set according to the task allocation benchmark analyzed in Sect. 6.1 where the number of agents was 100. The city was represented by a rectangular region of the Euclidean plane of size 10 × 10 km, divided into 25 neighborhoods of size 2 × 2 km, and 100 neighborhoods of size 1 × 1 km in settings 1 and 2, respectively. The setup includes 8-h shifts (as in real police departments), with 25 and 100 agents patrolling (one in each neighborhood) at the beginning of each shift in settings 1 and 2, respectively. The number of tasks arriving (i.e., the load) in a shift varied between 56, 111, 167, 222, and 278. Tasks arrived at a fixed rate and were uniformly distributed in random locations in the city.

The key differences between LEP benchmark and the task allocation benchmark are summarized in Table 6.6.

The parameters used for the simulation are presented in Table 6.7. These included four types of tasks of decreasing importance $I(\Psi) = 2400, 1600, 1200, 800$ from type 1 to type 4, respectively. Patrols had $I(\Psi) = 500$. Task types were selected randomly according to the distribution of real event types provided by the law enforcement authorities: 30, 40, 15, 15% of tasks were of types 1–4, respectively. The workloads were drawn from exponential distributions with mean values of 58, 55, 45, 37 for tasks of types 1–4, respectively [1]. The workloads are described using vectors of size two that represent the required workload for each skill. The workload for tasks of type 1 was $\vec{w}(\Psi_i) = (2w_i/3, w_i/3)$, for tasks of type 2 was $\vec{w}(\Psi_i) = (w_i/2, w_i/2)$, and for tasks of types 3 and 4 was $\vec{w}(\Psi_i) = (w_i)$. The *Cap* function varied for different types of tasks. The execution quality was monotonically non-decreasing as

Table 6.6 Key differences between law enforcement and task allocation problems

LEP	Task allocation
Dynamic agents	Static agents
Requires multiple different skills to perform a task	Requires specific skill to perform a task
Each agent possesses several unique skills	Each agent possesses one skill
Limited number of agents that can be allocated to a task	Unlimited number of agents that can be allocated to a task

Table 6.7 Summary of parameter values for simulation analysis

Parameter	Values
Area dimensions	10×10 km
Number of police officers (agents)	25, 100
Algorithms	FMC_TA^{H+}, HDBA, SA
Number of tasks	56, 111, 167, 222, 278
Number of shifts	100
Duration of shift	8 h
Importance of tasks	2400, 1600, 1200, 800
Importance of patrol	500
Distribution of tasks	30, 40, 15, 15%
Tasks location	Uniformly distributed at random
Control parameters	$\alpha = \beta = \gamma = 1$
Agents location	Initially 1 in each neighborhood

more agents perform the task concurrently, up to a specific number of agents for each of the skills.

When this specific number of agents for some skill is reached, adding additional agents with this skill can no longer improve execution quality. For tasks of types 3 and 4 the *Cap* function was:

$$Cap(\Psi_i, \vec{q}) = \begin{cases} I(\Psi) & \text{if } q_1 \geq 1, q_2 \in N \\ 0 & \text{otherwise} \end{cases} \tag{6.10}$$

For tasks of type 2 the *Cap* function was:

$$Cap(\Psi_i, \vec{q}) = \begin{cases} I(\Psi) & \text{if } q_1 \geq 1, q_2 \geq 1 \\ I(\Psi)/2 & \text{if } q_1 \geq 1, q_2 = 0 \vee q_1 = 0, q_2 \geq 1 \\ 0 & \text{if } q_1 = 0, q_2 = 0 \end{cases} \tag{6.11}$$

For tasks of type 1 the *Cap* function was:

$$Cap(\Psi_i, \vec{q}) = \begin{cases} I(\Psi) & \text{if } q_1 \geq 2, q_2 \geq 1 \\ I(\Psi)/2 & \text{if } q_1 \geq 2, q_2 = 0 \vee q_1 = 1, q_2 \geq 1 \\ I(\Psi)/4 & \text{if } q_1 = 1, q_2 = 0 \vee q_1 = 0, q_2 = 1 \\ 0 & \text{if } q_1 = 0, q_2 = 0 \end{cases} \tag{6.12}$$

The numerical computations were performed on a PC with 2.00 GHz CPU and 8 GB of RAM, using JAVA.

The algorithm's performance was compared for the measures of team utility, an average execution delay, percentage of abandoned tasks, percentage of shared tasks, and the average arrival time of agents to tasks.

For five different performance measures, Figs. 6.19, 6.20, 6.21, 6.22 and 6.23 graphically depict representative results of the simulations conducted to compare the studied algorithms as a function of shift load. Figure 6.19 shows, as expected, that at the lower shift loads of 55, 111, and 167 tasks, the team utility increases with the load for all three algorithms. However, in Fig. 6.19a at the higher loads of 222 and 278 tasks, the team utility continues to increase for FMC_TA^{H+} but decreases for HDBA and SA. For the high loads of 222 and 278 tasks, the difference in team utility between FMC_TA^{H+} and two other algorithms is significant ($p < 0.05$), with average team utility values for FMC_TA^{H+}, HDBA, and SA being 198,713.6, 129,849.1, and 134,892.9, respectively. In Fig. 6.19b at the higher loads of 222 and 278 tasks, the team utility continues to increase for HDBA and FMC_TA^{H+} but decreases for SA. For the high loads of 222 and 278 tasks, the difference in team utility between HDBA, FMC_TA^{H+}, and SA is significant ($p < 0.05$), with average team utility values for FMC_TA^{H+}, HDBA, and SA being 217,934.5, 230,713.5, and 136,719.3, respectively. Figure 6.19 shows that the average execution delay increases with shift loads for all three algorithms. In addition, in shifts with the highest load of 278

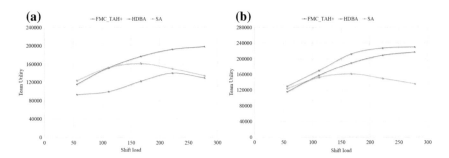

Fig. 6.19 Team utility as a function of shift load. **a** Set with 25 agents, **b** set with 100 agents

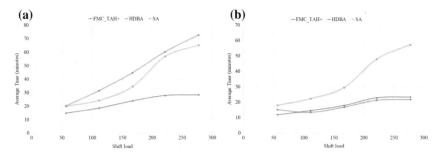

Fig. 6.20 Average arrival time to tasks as a function of shift load. **a** Set with 25 agents, **b** set with 100 agents

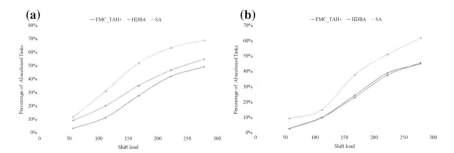

Fig. 6.21 Abandonment rate as a function of shift load. **a** Set with 25 agents, **b** set with 100 agents

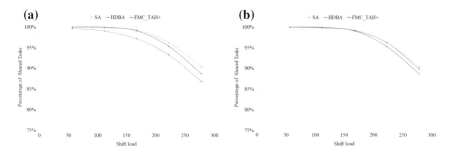

Fig. 6.22 Cooperation rate as a function of shift load. **a** Set with 25 agents, **b** set with 100 agents

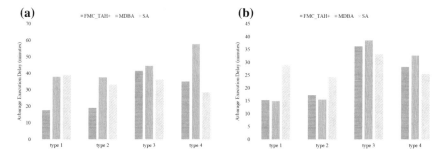

Fig. 6.23 Execution delay of tasks of different types for varying algorithms. **a** Set with 25 agents, **b** se with 100 agents

tasks, in Fig. 6.20a FMC_TA^{H+} has a lower average arrival time in minutes (28.29) compared to HDBA (72.74) and SA (65.31), and these differences are significant ($p < 0.05$); but in Fig. 6.20b HDBA has a lower average arrival time in minutes (21.81) compared to FMC_TA^{H+} (23.48) and SA (57.52), and these differences are significant ($p < 0.05$).

The measure depicted in Fig. 6.21 is the percentage of abandoned tasks, i.e., tasks where agents stopped execution before completion. According to Fig. 6.21, the percentage of abandoned tasks increases with the shift load for all three algorithms. At all shifts loads, FMC_TA^{H+} and HDBA have significantly lower percentages of abandoned tasks compared to SA ($p < 0.05$), indicating that FMC_TA^{H+} and HDBA finish significantly more tasks in comparison to SA. Figure 6.22 presents the percentage of shared tasks as a function of the shift load, showing above 95% task sharing at the lower shift loads of 55, 111, and 167 tasks for all three algorithms. However, for shifts with higher loads, the percentage of shared tasks decreases for all algorithms. At the highest shift load of 278, the statistically different percentages ($p < 0.05$) of shared tasks were 91.2, 88.9, and 87.3% for SA, FMC_TA^{H+}, and HDBA, respectively, in Fig. 6.22a and 90.3, 88.6, and 89.8% for SA, FMC_TA^{H+}, and HDBA, respectively, in Fig. 6.22b, indicating that all algorithms had a high collaboration rate during their execution.

Figure 6.23a shows that FMC_TA^{H+} successfully prioritizes important tasks, minimizing the execution delay in minutes to 17.6 and 19.1 for tasks of types 1 and 2, respectively ($p < 0.05$). However, the delays for tasks of types 3 (36.2) and 4 (28.5) were slightly smaller for SA with roughly the same execution delay for all task types regardless of the task importance. HDBA had the second-best execution delay (37.9) for tasks of type 1. The inherent preference of the high importance tasks by HDBA resulted in the worst execution delay of 57.7 min for the lowest importance tasks of type 4. Figure 6.23b illustrates that HDBA prioritizes important tasks more than FMC_TA^{H+} does, with execution delay of 14.9 and 15.5 min for tasks of types 1 and 2, respectively ($p < 0.05$). However, the delays for tasks of types 3 (36.2) and 4 (28.5) were slightly smaller for SA with roughly the same execution delay for all task types regardless of the task importance.

The described evaluation is shown to be effective in allocating dynamic tasks to heterogeneous police agents using simulation analyses. All algorithms are shown to solve the dynamic task allocation problem. For 100 agents, HDBA generated more collaboration among agents and resulted in better performance in terms of team utility and execution delay than both FMC_TA^{H+} and SA, with statistically significant 6 and 41% higher team utility in the highest shift load compared to FMC_TA^{H+} and SA, respectively.

One of the limitations of HDBA's performance in the first set was the low number of agents. Although small size swarms are able to cope with combinatorial optimization problems, a swarm in the order of 50 agents and above is usually needed to perform effectively on standard benchmarks [6]. An increase of the number of agents in the swarm from 50 to 100 can significantly increase the task allocation [35] and convergence to high quality solutions [25]. HDBA resulted in improved performance, achieving the best team utility in the second setup with 100 agents.

HDBA was shown to solve this dynamic task allocation benchmark, illustrating its ability to be applied for a different problem without any adaptations and parameter optimizations.

HDBA can be further improved by optimizing its parameters for this benchmark, which is one of the interesting future research directions. This must include the optimization of the control parameters α, β, and γ that provide a mechanism to adjust the swarm behavior and bias the importance of the urgency, distance, and skills, respectively. The optimization of these parameters can result in the reduction of the execution delay of important tasks of types 1 and 2, improving the overall utility.

References

1. Amador S, Okamoto S, Zivan R (2014) Dynamic multi-agent task allocation with spatial and temporal constraints. In: Proceedings of the 2014 international conference on autonomous agents and multi-agent systems, pp 1495–1496
2. Angeles R (2005) RFID technologies: supply-chain applications and implementation issues. Inf Syst Manage 22(1):51
3. Attaran M (2007) RFID: an enabler of supply chain operations. Supply Chain Manag Int J 12(4):249–257
4. Banks J, Hanny M, Pachano A, Thompson G (2007) RFID applied. Wiley, New Jersey
5. Bian F, Kempe D, Govindan R (2006) Utility based sensor selection. In: Proceedings of the 5th international conference on information processing in sensor networks, pp 11–18
6. Bratton D, Kennedy J (2007) Defining a standard for particle swarm optimization. In: IEEE swarm intelligence symposium, April, pp 120–127
7. Brintrup A, Ranasinghe D, McFarlane D (2010) RFID opportunity for leaner manufacturing. Int J Prod Res 48(9):2745–2764
8. Byers J, Nasser G (2000) Utility-based decision-making in wireless sensor networks. In: Proceedings of the 1st ACM international symposium on mobile ad hoc networking & computing. IEEE Press, Boston, MA, USA, pp 143–144
9. Closs DJ, McGarrell EF (2004) Enhancing security throughout the supply chain. IBM Center for the Business of Government, Washington, DC, pp 10–12
10. Dias EM, Fontana CF, Mori FH, Facioli LP, Zancul PJ (2008) Security supply chain. In: Mastorakis NE, Mladenov V, Bojkovic Z, Simian D, Kartalopoulos S, Varonides A (eds) WSEAS international conference on mathematics and computers in science and engineering, p 12
11. Dorigo M, Maniezzo V, Colorni A (1996) Ant system: optimization by a colony of cooperating agents. IEEE Trans Syst Man Cybern B Cybern 26(1):29–41
12. Eyers DR, Potter AT, Wang Y (2011) Supply chain implications of e-commerce channels for additive manufacturing. In: International conference of production research, Stuttgart, Germany
13. Floerkemeier C, Lampe M (2005) RFID middleware design—addressing both application needs and RFID constraints. GI Jahrestagung 1:277–281
14. Fokum DT, Frost VS, DePardo D (2009) Experiences from a transportation security sensor network field trial. Technical report ITTC-FY2009-TR-41420-11. The University of Kansas
15. Gao X, Xiang Z, Wang H, Shen J, Huang J, Song S (2004) An approach to security and privacy of RFID system for supply chain. In: IEEE international conference on e-commerce technology for dynamic e-business, pp 164–168
16. Gerkey BP, Matarić MJ (2003) Multi-robot task allocation: analyzing the complexity and optimality of key architectures. In: IEEE international conference on robotics and automation, pp 3862–3868
17. Gutin G, Punnen AP (eds) (2006) The traveling salesman problem and its variations, vol 12. Springer Science & Business Media
18. Halim AH, Ismail I (2019) Combinatorial optimization: comparison of heuristic algorithms in travelling salesman problem. Arch Comput Methods Eng 26(2):367–380
19. Haroun SA, Jamal B (2015) A performance comparison of GA and ACO applied to TSP. Int J Comput Appl 117(20)
20. Hore S, Chatterjee A, Dewanji A (2018) Improving variable neighborhood search to solve the traveling salesman problem. Appl Soft Comput 68:83–91

21. Ito T, Hattori H, Klein M (2007) Multi-issue negotiation protocol for agents: exploring nonlinear utility spaces. IJCAI 7:1347–1352
22. Ivanov D, Dolgui A, Sokolov B (2013) Multi-disciplinary analysis of interfaces "supply chain event management—RFID—control theory". Int J Integr Supply Manag 8(1/2/3):52–66
23. Jeong W, Nof SY (2009) A collaborative sensor network middleware for automated production systems. Int J Comput Ind Eng 57:106–113
24. Jevtić A, Gutiérrez A (2011) Distributed bees algorithm parameters optimization for a cost efficient task allocation in swarms of robots. Sensors 11(11):10880–10893
25. Karaboga D, Basturk B (2008) On the performance of artificial bee colony (ABC) algorithm. Appl Soft Comput 8(1):687–697
26. Ko HS, Azambuja M, Lee HF (2016) Cloud-based materials tracking system prototype integrated with radio frequency identification tagging technology. Autom Constr 63:144–154
27. Ko HS, Yoon S, Nof SY (2011) Intelligent alert systems for error and conflict allocation in supply networks. In: 18th IFAC world congress, Milano, Italy
28. Kongkaew W, Pichitlamken J (2012) A Gaussian process regression model for the traveling salesman problem. J Comput Sci 8(10):1749–1758
29. Lau HC, Zhang L (2003) Task allocation via multi-agent coalition formation: taxonomy, algorithms and complexity. In: Proceedings of the 15th IEEE international conference on tools with artificial intelligence, Sacramento, CA, USA, pp 346–350
30. Lawson B, Potter A, Pil FK, Holweg M (2019) Supply chain disruptions: the influence of industry and geography on firm reaction speed. Int J Oper Prod Manag
31. Lučić P, Teodorović D (2003) Computing with bees: attacking complex transportation engineering problems. Int J Artif Intell Tools 12(03):375–394
32. Lučić P, Teodorović D (2002) Transportation modeling: an artificial life approach. In: 14th IEEE international conference on tools with artificial intelligence, pp 216–223
33. Lánská M (2012) Supply chain security. In: Proceedings of the 9th international conference on logistics and sustainable transport, University of Maribor, Faculty of Logistics, Celje, pp 191–195
34. Mari SI, Lee YH, Memon MS, Park YS, Kim M (2015) Adaptivity of complex network topologies for designing resilient supply chain networks. Int J Ind Eng 22(1):102–116
35. Meng Y, Kazeem O, Muller JC (2007) A hybrid ACO/PSO control algorithm for distributed swarm robots. In: IEEE swarm intelligence symposium, April, pp 273–280
36. Modrák V, Moskvich V (2012) Impacts of RFID implementation on cost structure in networked manufacturing. Int J Prod Res 50(14):3847–3859
37. Naskar S, Basu P, Sen AK (2019) A literature review of the emerging field of IoT using RFID and its applications in supply chain management. In: Securing the internet of things: concepts, methodologies, tools, and applications. IGI Global, pp 1664–1689
38. Nelke SA (2016) Market equilibrium-based mechanism for dynamic task allocation. Doctoral book, Ben-Gurion University of the Negev, Faculty of Engineering Sciences, Department of Industrial Engineering and Management
39. Ngai EWT, To CK, Moon KK, Chan LK, Yeung PK, Lee MC (2010) RFID systems implementation: a comprehensive framework and a case study. Int J Prod Res 48(9):2583–2612
40. Park K, Min H, Min S (2016) Inter-relationship among risk taking propensity, supply chain security practices, and supply chain disruption occurrence. J Purch Supply Manag 22(2):120–130
41. Qiu RG (2007) RFID-enabled automation in support of factory integration. Int J Robot Comput Integr Manuf 23:677–683
42. Rana OF, Stout K (2000) What is scalability in multi-agent systems? In: Proceedings of the fourth international conference on autonomous agents. ACM, pp 56–63
43. Reinelt G (1991) TSPLIB—a traveling salesman problem library. ORSA J Comput 3:376–384
44. Rowaihy H, Eswaran S, Johnson M, Verma D, Bar-Noy A, Brown T, Porta TL (2007) A survey of sensor selection schemes in wireless sensor networks. In: Proceedings of SPIE, vol 6562
45. Sharma SK, Vasant BS (2015) Developing a framework for analyzing global supply chain security. IUP J Supply Chain Manag 12(3):7

46. Sidorov M, Ong MT, Sridharan RV, Nakamura J, Ohmura R, Khor JH (2019) "Ultralightweight mutual authentication RFID protocol for blockchain enabled supply chains." IEEE Access 7, 7273–7285

47. Spector L (2004) Automatic quantum computer programming: a genetic programming approach, vol 7. Springer Science & Business Media

48. Subashini S, Kavitha V (2011) A survey on security issues in service delivery models of cloud computing. J Netw Comput Appl 34(1):1–11

49. Tkach I, Edan Y, Nof SY (2012) Security of supply chains by automatic multi-agents collaboration. Inf Control Probl Manuf 14(1):475–480

50. Tkach I, Edan Y, Nof SY (2017) Multi-sensor task allocation framework for supply networks security using task administration protocols. Int J Prod Res 55:5202–5224

51. Tkach I, Jevtić A, Nof SY, Edan Y (2013) Automatic multi-sensor task allocation using modified distributed bees algorithm. In: IEEE international conference on systems, man, and cybernetics (SMC), Manchester, England, pp 1401–1406

52. Tkach I, Jevtić A, Edan Y, Nof SY (2018) A modified distributed bees algorithm for multi-sensor task allocation. Sensors 18(3):759. https://doi.org/10.3390/s18030759

53. Vu DM, Hewitt M, Boland N, Savelsbergh M (2019) Dynamic discretization discovery for solving the time-dependent traveling salesman problem with time windows. Transp Sci

54. Weinstein R (2005) RFID: a technical overview and its application to the enterprise. IT Prof 7(3):27–33

55. Whipple JM, Voss MD, Closs DJ (2009) Supply chain security practices in the food industry: do firms operating globally and domestically differ? Int J Phys Distrib Logist Manag 39(7):574–594

56. Williams NP, Liu Y, Nof SY (2002) TestLAN approach and protocols for the integration of distributed assembly and test networks. Int J Prod Res 40(17):4505–4522

57. Wolski R, Plank JS, Brevik J, Bryan T (2001) Analyzing market-based resource allocation strategies for the computational grid. Int J High Perform Comput Appl 15(3):258–281

58. Xiao F, Wang Z, Ye N, Wang R, Li XY (2018) One more tag enables fine-grained RFID localization and tracking. IEEE/ACM Trans Netw 26(1):161–174

59. Yoon Y, Kim YH (2013) An efficient genetic algorithm for maximum coverage deployment in wireless sensor networks. IEEE Trans Cybern 43(5):1473–1483

Chapter 7
Dual-Layer Multi-sensor Task Allocation System

This chapter presents an analysis of algorithms and protocols for sensor allocation in a dual-layer system. The dual-layer system framework is described, task administration protocols are presented, and the system with TAPs and HDBA that is able to handle the scenarios of deception, overloading, and tampering is defined in the current chapter. Figure 7.1 illustrates the dual-layer system architecture.

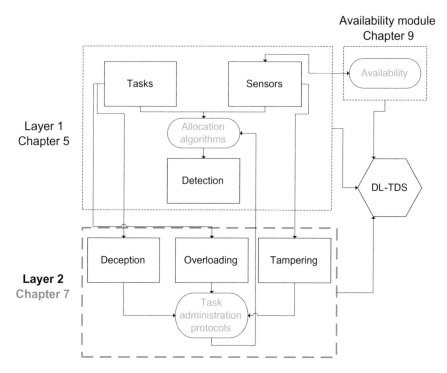

Fig. 7.1 Layer 2 in system architecture

© Springer Nature Switzerland AG 2020
I. Tkach and Y. Edan, *Distributed Heterogeneous Multi Sensor Task Allocation Systems*,
Automation, Collaboration, & E-Services 7,
https://doi.org/10.1007/978-3-030-34735-2_7

Despite the extensive ground covered by researchers thus far regarding multi-agent task allocation algorithms, several serious shortcomings in the literature have been found and must be addressed. Various authors have outlined frameworks based on design principles derived from observing real systems; however, no framework has been proposed to deal with heterogeneous sensors to multiple tasks without knowing a priori when, where, and which task will enter or leave the environment and handle problems within the sensory system.

Recent research in multi-agent task allocation methods and meta heuristics has provided a plethora of different algorithms; nevertheless, there is a lack of understanding on how the problems in sensors and in the allocation process such as very long attendance times and many tasks with same priorities can be handled and optimally addressed. Furthermore, studies have addressed agent-level and administration-level allocation in isolation, neglecting their interrelation and the benefits of combining them into a dual-layer system. Not only these levels have been addressed in isolation, but they also cannot fully handle problems in the sensory system. Previous research proposed task administration protocols to deal with three elements in task administration: task, resource, and process. This book extends and modifies these protocols to deal with a dual-layer multi-sensor task allocation system and to analyze the status of sensors to modify the allocation if necessary.

7.1   Definitions

In order to overcome uncertainties in task arrival and sensory performance and disturbances, a dual-layer approach is required. The dual-layer system is responsible for actively making decisions and triggering timely actions so that the overall system performance can be further improved. It can handle problems existing in allocation methods and algorithms due to their inherent limitations. In order to overcome such limitations, protocols within this dual-layer system must repeatedly identify the current state of the system and take proper actions to deal with allocation problems. These problems include:

1. Tasks that may have much higher priority over other tasks; this will occupy the sensors without the ability to perform other tasks that are close to their deadline and must be handled quickly before less urgent tasks.
2. Many tasks have the same priorities; the system may need to reprioritize them to avoid conflicts in sensor allocation.
3. Tasks that require long execution times but have low priority may occupy the sensors.
4. A failure of a portion of sensors in the system, which may affect the allocation process.

These problems are mapped in relation to the allocation and sensors limitations:

1. **Overloading**: Tasks that require very long attendance times. These tasks may occupy the sensors that are allocated to them for long periods of time thereby delaying sensors from handling other tasks. The sensors are overloaded with a portion of tasks that delay them. A time out policy that recognizes if a sensor is experiencing a delay while other tasks are waiting for execution, can overcome this problem.
2. **Deception**: Sensors are occupied by some tasks that may be neither urgent nor important and be delayed by them. This can be caused by tasks that may have higher priority over other tasks or many tasks with similar priorities. Due to malfunction or recognition problems, tasks may be perceived by the system with a higher priority than they should. These tasks may unnecessarily occupy the sensors. Thus reducing sensors' ability to detect other tasks that are close to their deadline and must be handled quickly before less urgent tasks. These cases require the system to reprioritize tasks.
3. **Tampering**: Failure of a portion of sensors in the system. Sensors may fail due to internal properties (i.e., reliability of the hardware) and external reactions (i.e., weather conditions, jamming). The system should be able to monitor this and ensure that system performance will not be decreased.

Definition 1 Task's deadline (Δ), task's initial completion time value t_{init}, time out value to, current time t_c, set of tasks in the system (M), task processing time and a queue of tasks;
Overloading (OVL) occurs when

$$OVL = \{M : t_{init} \gg 0 \& \Delta \gg 0 | (t_c - tst > to) \& (\text{queue}(M) > 0)\}$$

Definition 2 Perceived task's priority (F), calculated task's priority (T), and set of tasks in the system (M);
Deception (DCP) occurs when $DCP_1 = \{\forall a, b \in M : F(a) = F(b) | T(a) < T(b)\}$

$$DCP_2 = \{\forall a, b \in M : F(a) >> F(b) | T(a) < T(b)\}$$

Definition 3 Performance of a sensor (V), set of sensors in the system (N), and set of tasks in the system (M);
Tampering (TMP) occurs when $TMP = \{\forall k \in N : V_{ik} = 0 | V_{ik} \text{ allocated to } M\}$ for all i

7.2 Task Administration Protocols

Task Administration Protocols (TAPs) are defined as a set of protocols responsible for making decisions actively and triggering timely actions so that these decisions and triggers can improve performance [1–3, 5]. Previous research regarding TAPs dealt with three elements in task administration: task, resource, and process. According to Ko and Nof [5], TAPs were designed as a combination of three sub-protocols in order to handle three administration elements: task requirement analysis protocol (TRAP), shared resource allocation protocol (SRAP), and synchronization and time-out protocol (STOP). Each of the protocols is activated by the TAP mechanism upon a certain trigger and acts according to the appropriate administration element.

TAPs were used in the literature for task initialization, resource allocation, and process monitoring. TRAP was used to analyze and assign priorities to tasks inserted into a task queue. Considering the current status of resources, SRAP decides which resource a task must be assigned. STOP was used to control procedures in which a resource may be occupied excessively by a task, preventing other tasks from being processed by this resource. Beyond a certain threshold of resource occupation, the task needs to be returned to the queue so that other tasks can be processed at the resource.

According to Ko and Nof [5], different TAP schemes must be applied depending on the system structure and the type of resources and tasks at hand.

The design of TAPs for task administration in a collaborative production/service system was investigated by Ko and Nof [5]. In this application, TAPs were designed as a control mechanism that can manage complicated situations in the collaborative task workflow environment. A case study of applying TAPs for TestLAN, as an example of collaborative production systems, was developed by Williams et al. [9] and enhanced by Ko and Nof [5] to analyze the design of TAPs and show the advantage of TAPs over non-TAP coordination protocols. The simulation analysis indicated that TAPs performed significantly better than other non-TAPs, particularly under medium or high load conditions. The second case study analyzed by Ko and Nof [6] included the application of TAPs to FSN (Facility Sensor Network) as an example of collaborative service systems. FSN is a wireless sensor network to monitor processes in collaborative production facilities. In FSN the data are transmitted under a routing protocol, which tries to maximize the lifetime of the wireless sensor network [4], and then processed with a data processing protocol to obtain meaningful information from the data. Their simulation results indicated that the performance under TAPs was significantly better than under simple coordination protocols.

These studies have addressed agent-level and administration-level allocation in isolation, neglecting their interrelation and the benefits of combining them into a dual-layer system. This chapter combines an agent-level task allocation algorithm with task administration protocols with an addition of a new protocol for analyzing the status of agents to modify the allocation if necessary.

Task administration protocols (TAPs) are used to overcome uncertainties in task arrival and sensory performance and disturbances (i.e., sensor failure, conflicts in task priorities, high time-consuming tasks) in the multi-sensor system (Fig. 7.2). TAPs consist of four protocols, one of the protocols applies a bio-inspired Heterogeneous Distributed Bees Algorithm (HDBA) for sensors allocation and real time allocation of tasks. This algorithm efficiently allocates a high number of sensors to upcoming tasks in order to handle as many tasks as fast as possible.

Optimal sensor's availability is related to their monetary cost in the system achieved by deployment of redundant sensors (see Chap. 10). Employing TAPs that use HDBA allows dynamic, real time allocation of distributed sensors to tasks when they occur.

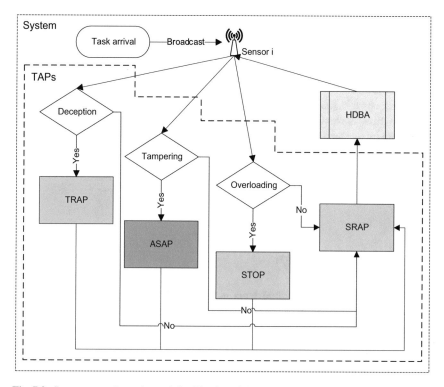

Fig. 7.2 System procedure scheme (after Tkach et al. [7])

In the following situations, another layer may be required to handle problems in the allocation process. These situations are mapped in relation to the allocation function and sensors limitations:

1. Tasks that require very long execution time (7.1) may occupy the sensors that are allocated to them for long periods of time and decrease system performance dramatically by delaying these sensors without the ability to handle other tasks.

$$OVL = \{M : t_{init} \gg 0 \& \Delta \gg 0 | (t_c - tst > to) \& (\text{queue}(M) > 0)\} \qquad (7.1)$$

A time out policy that recognizes if a sensor is experiencing a delay while other tasks are waiting for execution, can overcome this problem.

2. A batch of tasks that may have much higher priority over other tasks or many tasks that have the same priority (7.2, 7.3)

$$DCP_1 = \{\forall a, b \in M : F(a) = F(b) | T(a) < T(b)\} \qquad (7.2)$$

$$DCP_2 = \{\forall a, b \in M : F(a) >> F(b) | T(a) < T(b)\} \qquad (7.3)$$

A similar problem may arise in cases where tasks are perceived by the system with a higher priority than it should be, due to malfunction or recognition problems.

Those tasks may unnecessarily occupy the sensors without the ability to perform other tasks that are close to their deadline and need to be handled quickly before less urgent tasks; in these cases, the system may need to reprioritize them.

3. A failure of a portion of sensors in the system defined as tampering (7.4)

$$TMP = \{\forall k \in N : V_{ik} = 0 | V_{ik} \text{ allocated to } M\} \text{ for all } i \qquad (7.4)$$

7.3 Framework Design

The system is designed as a dual-layer network; a process layer and a monitoring layer (Fig. 7.3). The process layer consists of multiple sensors and is responsible for allocating them to complete tasks. As multi-sensor systems are vulnerable to some risks and problems [8], the monitoring layer functions at a higher level than the process layer and is used to monitor those problems in the process layer, applying TAPs to handle them. The problems in the process layer are defined as overloading, deception, and tampering sensors.

The system executes task administration protocols to ensure efficient operation by the system executing the following four activities [7] (Fig. 7.4):

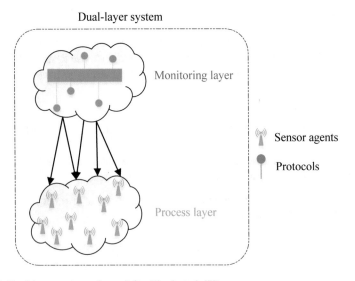

Fig. 7.3 Dual-layer system scheme (after Tkach et al. [7])

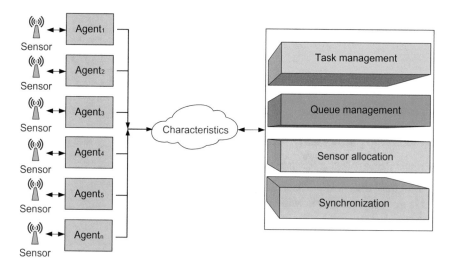

Fig. 7.4 Multi-sensor collaboration model (after Tkach et al. [7])

A1. **Task managing**: decides how to reassign the priorities of the required tasks based on characteristics of each task.

A2. **Queue managing**: manages tasks queue such that no task will ever be delayed. This activity is important since delaying a task can cause it to lose its relevance.

A3. **Sensor allocation**: executes sensor allocation algorithm for the sensor to task allocation. The sensor selection determines which sensors are appropriate to perform the task in order to optimize the performance of the overall system. This activity is responsible for executing the process layer, which uses HDBA for sensors to task allocation.

A4. **Synchronization**: sets up a timeline of commands for every sensor to carry out tasks based on their availability and performances, given task requests. It ensures that tasks are not unnecessarily delayed while a current task is experiencing delays, given that there are other tasks waiting in the queue.

7.4 Task Administration Protocols

Four task administration protocols (TAPs) are implemented for PL and ML following Ko and Nof [5, 6] and based on the four activities above [7] (Table 7.1):

"TRAP" (task requirement analysis protocol):

Reassigns priority levels to monitoring and allocation tasks so that they may be sequenced in the queue.

"ASAP" (assignment analysis protocol):

It decides if the task should wait for the backup sensors to become active or be processed with a set of available functional sensors.

In "ASAP" there are two options:

Table 7.1 Summary of protocols and their characteristics

Admin element	Task	Space	Resource	Time
Protocol	TRAP	ASAP	SRAP	STOP
Related event	Task arrival	Sensors analysis	Sensors selection	Excessive process time
Purpose	Prioritize tasks	Analyze assignment	Effective allocation	Synchronization
Action	Identify task requirements	Identify spatial limitations	Decentralized decision making	Treat uncertainty in processes

– Wait until the backup resources are activated.
– Do not wait and operate without the best allocation.

"SRAP" (sensors allocation protocol):
Its function is to consider the priority level assigned to a given task relative to the availability of sensors and then allocate the task to appropriate sensors based on its priority using HDBA.

"STOP" (synchronization and time-out protocol):
It is activated to monitor if the current task in service needs to be timed-out because of its excessive use after a given period. The objective is to ensure that no sensors are unnecessarily delayed.

The steps for implementing these protocols are:

Step 1. Identify the new task arrival to the system.
Step 2. Task sequenced in the queue according to the highest priority and earliest arrival time.
Step 3. If the benefit from processing the task with the default allocation of available sensors without activating backup sensors is higher than the benefit from the collaboration of an optimal set of sensors minus penalty for waiting then execute a task, remove the task from queue, and update queue.
Step 4. Sensors availability is checked and the sensor is allocated.
Step 5. Sends a message that the current task processing is completed.
Step 6. The highest priority task with the earliest arrival time is selected from the head of the queue to be processed next.

7.5　TAPs Implementation

The task priority is assigned by Eq. (7.5) and considers the following conditions [7]:

(1) The relatively higher Ψ_i's profit (value) is, the higher its priority.
(2) If Ψ_i requires the same allocation as the current task, its priority is high since setup time is minimized or not needed.
(3) If the current task is not yet finished but is close to being completed, it has a relatively higher priority.
(4) A task relatively closer to its deadline has a higher priority.

(5) The earlier the task's arrival time the higher its priority.

$$T_i = \eta_i \times \left(f_i + \frac{1}{(tct_i - t_{current})} + \frac{1}{t_{arrival\ i}} \right) \tag{7.5}$$

where T_i is a calculated value of the new priority of a task according to protocol TRAP, i is the task index, f_i is the initial priority value of task i, η is the assignment type request, tct is the task completion time, $t_{current}$ is the current time, and $t_{arrival}$ is the task arrival time.

$$\eta = \begin{cases} 1 & if\ same\ requested\ assignment \\ 0.5 & otherwise \end{cases} \tag{7.6}$$

The decision regarding *wait for backup sensors* or *proceed with the task autonomously* is performed by calculating the difference between the performance of the system in default allocation and the optimal set of sensors, and the penalty for the waiting times (7.7).

$$Q_S = V_{IS_{default}} - V_{OS} + V_t \times \tau \tag{7.7}$$

where, Q_S is the assignment analysis function, $V_{ISdefault}$ is the performance of default allocated sensors without activating backup sensors, V_{OS} is the performance of the optimal set of sensors with backup sensors, V_t is waiting time penalty value, and τ is the waiting time in seconds.

A possible solution for sensor selection and allocation can be derived using the HDBA. The stop criterion is given by a threshold that quantifies the desired performance of the sensors for a given task.

Time-out value is given by (7.8).

$$to = \mu_{ct} + 2\sigma_{ct} \tag{7.8}$$

where μ is task mean completion time, σ is the standard deviation of task completion time.

Once a job is timed-out, its arrival time is modified to reflect the time when it re-enters the system, and all the service and waiting times must be updated.

For the "STOP" implementation it is necessary to monitor the time a task has been in process. If a task's service time is greater than its time-out period and there are tasks in the queue, the current job is timed-out and returns to the queue with an identifier indicating its priority level and remaining process time. The exceptions for "STOP" are:

1. Task processing time has not exceeded the time-out value *to*.
2. No other task is waiting in the queue. If queue $= 0$ then continue task processing.
3. The task is close to completion. If $tct\text{-}t_{current} \le \theta$ then continue task processing.

$$STOP = \begin{cases} false & \begin{aligned} & \text{if}(t_{current}\, tst_i \leq to) \text{ or } (\text{queue} = 0) \text{ or} \\ & \text{or } (tct_i - t_{current} \leq \theta) \end{aligned} \\ true & otherwise \end{cases} \quad (7.9)$$

where, θ is a threshold, tst_i is task start time.

Pseudocode of the dual-layer system

1. Initialize:
   Set $t=0$
   Place N agents 'bees' on sensors
2. If *termination condition met* then go to step 6
   else
       Upon task arrival calculate the task priority f_i
       Activate TRAP
           Calculate the new task priority function T_i
           Update queue of tasks based on their new priorities
       Go to step 3
3. If *new task arrived* then go to step 2
   If there are failed sensors
       Activate ASAP
       Calculate Q_S
       If $Q_S>0$ then make a default assignment and go to step 4
       else
       Activate backup sensors and go to step 4
4. For $i=1$ to M do
       For $k=1$ to N do
           Calculate probabilities for each sensor p_{ik}
       Apply wheel-selection rule
   Allocate sensors according to the selection and go to step 5
5. If $(t_{ex_i}=0)$ then set $\Delta_i=0$ and go to step 2
   else
       Calculate to
       If STOP is *true* then set $\Delta_i=0$ and go to step 2
       else
           Go to step 4
6. Finalize:
       Calculate V_l
       Calculate T
       Stop

References

1. Anussornnitisarn P, Peralta J, Nof SY (2002) Time-out protocol for task allocation in multi-agent systems. J Intell Manuf 13(6):511–522
2. Huang CY, Nof SY (2000) Formation of autonomous agent networks for manufacturing systems. Int J Prod Res 38(3):607–624
3. Huang CY, Nof SY (2002) Evaluation of agent-based manufacturing systems based on a parallel simulator. Comput Ind Eng 43(3):529–552
4. Jeong W, Nof SY (2008) Performance evaluation of wireless sensor network protocols for industrial applications. J Intell Manuf 19(3):335–345
5. Ko HS, Nof SY (2010) Design of protocols for task administration in collaborative production systems. Int J Comput Commun Control 5(1):91–105
6. Ko HS, Nof SY (2012) Design and application of task administration protocols for collaborative production and service systems. Int J Prod Econ 135(1):177–189
7. Tkach I, Edan Y, Nof SY (2017) Multi-sensor task allocation framework for supply networks security using task administration protocols. Int J Prod Res 55:5202–5224
8. Walters JP, Liang Z, Shi W, Chaudhary V (2007) Wireless sensor network security: a survey. Secur Distrib, Grid, Mob, Pervasive Comput 1:367
9. Williams NP, Liu Y, Nof SY (2002) TestLAN approach and protocols for the integration of distributed assembly and test networks. Int J Prod Res 40(17):4505–4522

Chapter 8
Extended Example of Dual-Layer Multi-sensor Task Allocation Systems

This chapter describes an example of a multi-sensory security system for supply networks with disturbances. This example is evaluated by the use of a dual-layer system to illustrate the methods described in Chap. 7.

The purpose of adding a second layer is to handle problems of deception, overloading, and tampering in the allocation process. The dual-layer system employs the advantages of HDBA with TAPs. Due to its decentralized heuristic nature, HDBA cannot guarantee an optimal solution, but it can provide sub-optimal solutions to NP-hard problems. By using TAPs these solutions were found to be further improved.

The dual-layer system can be applied to many different types of applications including LEP. Once the problems of deception, overloading, and tampering are introduced to LEP, adding a second layer will increase the performance (similar to the task allocation benchmark presented in this chapter). This can be studied as part of future research.

8.1 Dual-Layer System Evaluation

A dual-layer sensory system is illustrated in Fig. 8.1. Such a sensory system must be able to overcome several problems. The first problem has to do with tasks that require very long attendance times. These tasks may occupy the sensors allocated to them for long time durations, thus delaying the handling of other tasks by those sensors. In this overloading problem, the sensors are overloaded with a portion of tasks that delay them. A time-out policy, which recognizes if a sensor is experiencing a delay while other tasks are waiting for execution, can overcome this problem. The second problem that the sensory system must overcome relates to tasks that may have much higher priority over other tasks or the same priority as other tasks. A similar problem may arise when, due to malfunction or recognition problems, tasks are perceived by the system to be of a higher priority than they should be. Such tasks may unnecessarily occupy some sensors. The system may need to reprioritize those sensors, which are unable to perform other tasks that must be handled quickly,

© Springer Nature Switzerland AG 2020
I. Tkach and Y. Edan, *Distributed Heterogeneous Multi Sensor Task Allocation Systems*,
Automation, Collaboration, & E-Services 7,
https://doi.org/10.1007/978-3-030-34735-2_8

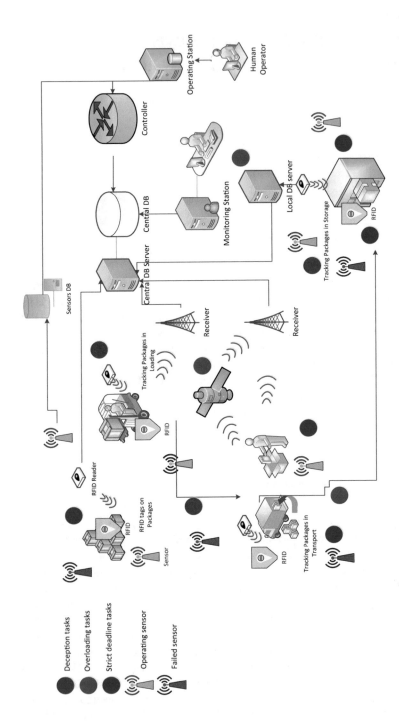

Fig. 8.1 An illustration of a dual-layer logistic system with distributed sensors to monitor security. Deception is marked by red circles, overloading by blue circles, and important tasks by purple circles

ahead of less urgent tasks, due to being close to their deadline. In this deception problem, sensors are occupied and delayed by some tasks that may be neither urgent nor important. The third problem that the sensory system must overcome is related to a failure of a portion of sensors in the system. In this tampering problem, sensors may fail due to internal properties (e.g., hardware reliability) and external factors (e.g., weather conditions). The system should be able to monitor these problems to ensure that system performance is unharmed.

The simulation compares the performance of a dual-layer system incorporating TAPs to a single-layer task allocation system without TAPs. The parameters used for the simulation are presented in Table 8.1.

Tasks were uniformly distributed in the arena based on Rowaihy et al. [2] and Byers and Nasser [1]. The sensors' range values were predefined to cover 15 to 45 meters relative to the size of the arena (100×100 m). An inherent noise in each sensor was introduced according to typical operational distance ranges of known commercial sensors [3] relative to the arena size: acoustic sensor, seismic sensor, and forward-looking infrared radar (FLIR). The allocation range of these sensors is given by:

$$R_{seismic} = 0.8 \times R_{acoustic} = 1.25 \times R_{FLIR}$$

The numerical computations were performed on a PC with 2.90 GHz CPU and 12 GB of RAM, using Matlab R2015a software. The following performance measures were analyzed:

Table 8.1 Summary of parameter values for simulation analysis

Parameter	Values
Area dimensions	100×100 m
Number of sensors ('bees')	100
Number of tasks	200
Task completion time	Randomly distributed from 0 to 10 min
Simulation duration	200 steps
Tasks' locations	Uniformly distributed at random
Control parameters	$\alpha = \beta = \gamma = 1$
Sensors' locations	Uniformly distributed random
Task's arrival time	Every 1 step
Sensor's range coverage	Predefined from 15 to 45 m
Number of false tasks	10, 20, 50
Number of failed sensors	10, 20, 30
Number of high-time consuming tasks	10, 20, 50
TAPs	TRAP, ASAP, SRAP, STOP

Table 8.2 Values of deception, tampering, and overloading for simulation analysis

	Deception (number of false tasks)	Fault tolerant/tampering (number of failed sensors)	Overloading (number of high time-consuming tasks)
	10	10	10
	20	20	20
	50	30	50
Responsible protocols	TRAP	ASAP	STOP

1. Number of treated tasks by each sensor is defined by: $\chi_k = \sum_{i=1}^{M} task_i;$ $\begin{array}{l} task_i = 1 \text{ if } V_{ik} > 0 \\ else\, task_i = 0 \end{array}$,
   where χ_k is the number of treated tasks of k-th sensor, M is the total number of tasks in the system, i is the index of the current task, and V_{ik} is the k-th sensor's performance on the i-th task.
2. Number of important tasks treated.
3. Number of unallocated tasks is defined by: $\varepsilon = \sum_{i=1}^{M} task_i;$ $\begin{array}{l} task_i = 1 \text{ if } \sum_{k=1}^{N} V_{ik} = 0 \\ else\, task_i = 0 \end{array}$,

where ε is the number of unallocated tasks, N is the total number of sensors in the system.

The simulation values for situations of overloading, deception, and tampering are summarized in Table 8.2. For each situation, an appropriate protocol was assigned.

The mean values obtained from 100 independent runs were compared at the statistical confidence level of 95%.

8.2 Results and Discussion

Graphs of the simulated results are presented in Figs. 8.2, 8.3, 8.4, 8.5, 8.6, 8.7, 8.8, 8.9 and 8.10 for scenarios of overloading, deception, and tampering. Each scenario was compared for three performance measures: the number of treated tasks by each sensor, the number of unallocated tasks, and the number of important tasks treated.

Figures 8.2, 8.3 and 8.4 illustrate the mean number of treated tasks by each sensor as a function of the number of false tasks, high time-consuming tasks, and failed sensors in single-layer and dual-layer systems. Results revealed that, as expected, as the number of false tasks increased (Fig. 8.2), the mean number of treated tasks by each sensor decreased (from 43.22 to 16.47 and from 43.21 to 34.78, in a single-layer system with HDBA, and dual-layer system using TAPs, respectively). Moreover, with the increase in the number of false tasks, the difference in the number of treated tasks

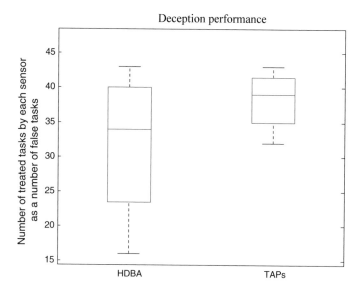

Fig. 8.2 Number of treated tasks by each sensor for a different number of false tasks. The bars represent mean values of 100 independent runs, the edges of the box are the 25th and 75th percentiles, the whiskers are the most extreme data points

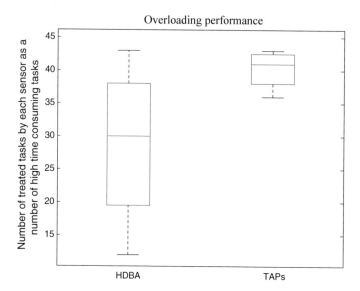

Fig. 8.3 Number of treated tasks by each sensor for a different number of high time-consuming tasks. The bars represent mean values of 100 independent runs, the edges of the box are the 25th and 75th percentiles, the whiskers are the most extreme data points

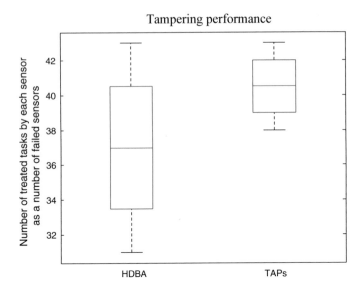

Fig. 8.4 Number of treated tasks by each sensor for a different number of failed sensors. The bars represent mean values of 100 independent runs, the edges of the box are the 25th and 75th percentiles, the whiskers are the most extreme data points

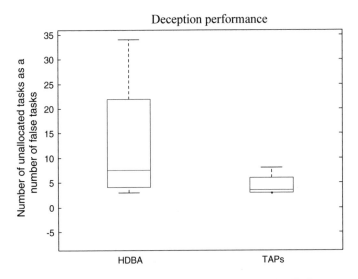

Fig. 8.5 Number of unallocated tasks for a different number of false tasks. The bars represent mean values of 100 independent runs, the edges of the box are the 25th and 75th percentiles, the whiskers are the most extreme data points

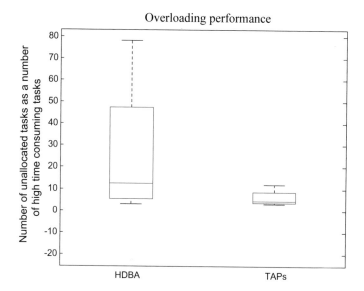

Fig. 8.6 Number of unallocated tasks for a different number of high time-consuming tasks. The bars represent mean values of 100 independent runs, the edges of the box are the 25th and 75th percentiles, the whiskers are the most extreme data points

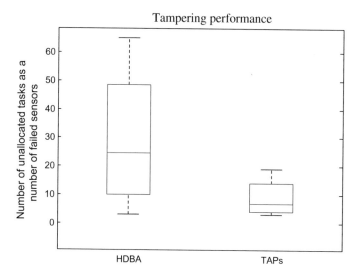

Fig. 8.7 Number of unallocated tasks for a different number of failed sensors. The central mark is the median value of 100 independent runs, the edges of the box are the 25th and 75th percentiles, the whiskers are the most extreme data points

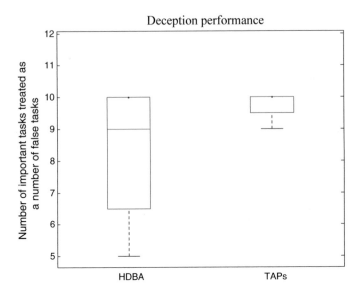

Fig. 8.8 Number of important tasks treated for a different number of false tasks. The bars represent mean values of 100 independent runs, the edges of the box are the 25th and 75th percentiles, the whiskers are the most extreme data points

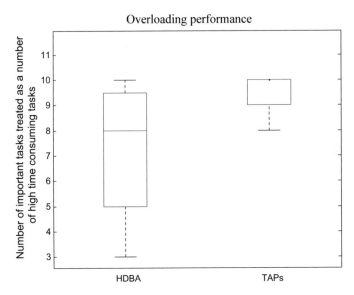

Fig. 8.9 Number of important tasks treated for a different number of high time-consuming tasks. The bars represent mean values of 100 independent runs, the edges of the box are the 25th and 75th percentiles, the whiskers are the most extreme data points

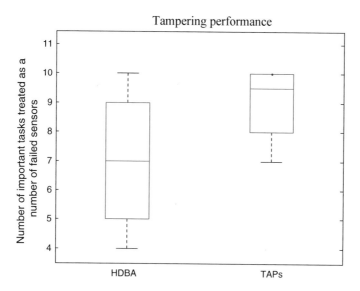

Fig. 8.10 Number of important tasks treated for a different number of failed sensors. The bars represent mean values of 100 independent runs, the edges of the box are the 25th and 75th percentiles, the whiskers are the most extreme data points

between TAPs and HDBA was also increased. This indicates that the monitoring layer performed effectively by minimizing the decrease in the number of treated tasks. Similarly, as the number of high time-consuming tasks (Fig. 8.3) and number of failed sensors (Fig. 8.4) increased, the mean number of treated tasks by each sensor also decreased and the difference in the number of treated tasks between TAPs and HDBA increased, respectively.

Figures 8.5, 8.6 and 8.7 illustrate the mean number of unallocated tasks as a function of the number of false tasks, high time-consuming tasks, and failed sensors in single-layer and dual-layer systems. The difference in the number of unallocated tasks between TAPs and HDBA was greatly increased in all scenarios. Thus, the dual-layer system managed to greatly reduce the number of unallocated tasks by more than 72% (from 31.07 to 8.58, 72.76 to 13.78, and 65.2 to 19.72 in deception, overloading, and tampering scenarios, respectively).

Figures 8.8, 8.9 and 8.10 illustrate the mean number of important tasks that were successfully treated by the system, as a function of the number of false tasks, high time-consuming tasks, and failed sensors in single-layer and dual-layer systems. In these scenarios, the number of treated important tasks was increased by more than 80% in the dual-layer system in comparison to the single-layer system (from 5.06 to 9.01, 3.03 to 8.07, and 4.02 to 7.01 in deception, overloading, and tampering scenarios, respectively). The simulation results are summarized in Table 8.3.

Table 8.3 Summary of results for scenarios of overloading, deception, and tampering with 100 independent runs

Measured value	Number of false tasks	Deception				Number of high time consuming tasks	Overloading				Number of failed sensors	Tampering			
		HDBA	Standard deviation	TAPs	Standard deviation		HDBA	Standard deviation	TAPs	Standard deviation		HDBA	Standard deviation	TAPs	Standard deviation
Number of treated tasks by each sensor	0	43.22ᵃ	8.09	43.21ᵃ	11.62	0	43.14ᵃ	10.59	43.47ᵃ	11.54	0	43.78ᵃ	4.55	43.74ᵃ	9.76
	10	38.47ᵃ	13.31	40.50ᵃ	8.14	10	33.58ᵃ	4.22	42.95ᵃ	4.26	10	38.38ᵃ	1.61	41.52ᵃ	5.33
	20	31.37ᵃ	2.12	39.44ᵃ	9.62	20	27.32ᵃ	8.36	40.62*	4.79	20	36.10ᵃ	7.93	40.13ᵃ	6.63
	50	16.47ᵃ	0.15	34.78ᵃ	5.59	50	13.65ᵃ	9.06	37.26ᵃ	2.23	30	31.45ᵃ	8.23	39.35ᵃ	2.79
Number of unallocated tasks	0	3.57	5.52	3.29	1.16	0	3.44	0.30	3.80	2.48	0	3.78	9.71	3.92	2.32
	10	5.84	3.52	3.26	3.87	10	9.99	6.45	5.90	4.74	10	18.35ᵃ	2.53	4.19ᵃ	0.62
	20	10.64	8.78	4.63	7.78	20	19.62ᵃ	12.20	8.55ᵃ	10.53	20	32.93ᵃ	3.22	8.09ᵃ	3.66
	50	31.07ᵃ	6.01	8.58ᵃ	2.04	50	72.76ᵃ	9.63	13.78ᵃ	7.36	30	65.20ᵃ	5.44	19.72ᵃ	10.65
Number of important tasks treated	0	10.71ᵃ	4.47	10.51ᵃ	13.42	0	10.66ᵃ	5.14	10.28ᵃ	3.30	0	10.23ᵃ	13.13	10.31ᵃ	3.22
	10	10.57ᵃ	10.17	10.08ᵃ	7.65	10	9.73ᵃ	4.19	10.26ᵃ	13.15	10	8.39ᵃ	2.08	10.61ᵃ	11.18
	20	8.81ᵃ	5.54	10.10ᵃ	10.91	20	7.04ᵃ	1.88	10.12ᵃ	5.32	20	6.98	11.38	9.83	5.27
	50	5.06	5.09	9.01	9.53	50	3.03	3.90	8.07	5.03	30	4.02	3.78	7.01	6.77

The first sub-column of each problem represents the result for each tested scenario using HDBA, the second sub-column represents the result for each tested scenario using TAPs

ᵃStatistically significant results

References

1. Byers J, Nasser G (2000) Utility-based decision-making in wireless sensor networks. In: Proceedings of the 1st ACM international symposium on Mobile ad hoc networking and computing, IEEE Press, Boston, MA, USA, pp 143–144
2. Rowaihy H, Eswaran S, Johnson M, Verma D, Bar-Noy A, Brown T, Porta TL (2007) A survey of sensor selection schemes in wireless sensor networks. In: Proceedings of SPIE, vol 6562
3. Yoon Y, Kim YH (2013) An efficient genetic algorithm for maximum coverage deployment in wireless sensor networks. IEEE Trans Cybernetics 43(5):1473–1483

Chapter 9
Fault Tolerant Multi Sensor System with High Availability

This chapter describes and analyzes the 'Availability module' of the task allocation system (Fig. 9.1). This chapter presents the reliability design and availability analysis of the system to further optimize system performance in case of sensor failure and to ensure fault tolerant allocation system performance. Availability optimization of the sensors' operation and comparison of the dual-layer system performance with

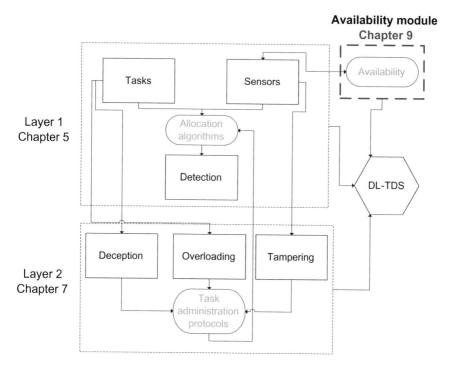

Fig. 9.1 Availability module in system architecture

© Springer Nature Switzerland AG 2020
I. Tkach and Y. Edan, *Distributed Heterogeneous Multi Sensor Task Allocation Systems*,
Automation, Collaboration, & E-Services 7,
https://doi.org/10.1007/978-3-030-34735-2_9

optimized sensors' availability to a regular non-optimized dual-layer system was conducted for two system types using Monte Carlo simulations.

9.1   System Availability

One of the important issues in systems design is improving the fault tolerance and resilience [8] of system operation. A multi-sensory system can be made fault-tolerant by designing redundancy into the system [6]. Some researchers suggest that the fault tolerance of the sensory system can be increased by putting redundant sensors to sleep and awakening these sleeping sensors when they are needed to restore task coverage [18]. This is an important characteristic since most applications in dynamic environments rely on continuous progress even if some system components fail (see [5]). For example, in [7] some monitoring sensors in automated production system middleware expected failures during task-execution, but it was still required that the overall mission be completed in the best way possible given the remaining resources. Therefore, real world applications in monitoring require a novel system in which some sensors are redundantly distributed with different sensing capabilities (i.e., heterogeneous sensors).

The availability of a system defines the percentage of time in which the system is operational out of the total system employment time. This is a critical value that influences the system uptime and its fault tolerance. In general, there are two numerical techniques for estimating and optimizing system availability: Monte Carlo simulation and numerical solution of the underlying Markov chain, in addition to an analytical solution that mainly focuses on non-shortage probability (NSP). Simulation is often the preferred solution technique for modeling real-world systems, since it is capable of solving large, complex systems with parallel components and general distribution functions. During Monte Carlo simulation, the systems' model is repeatedly executed, and a new random number sequence is used for each execution to compute a different failure value with each execution [1].

Reliability Block Diagram (RBD) is the most common method for expressing the reliability of systems [4]. RBD represents the reliability of a system by composing a set of blocks that are arranged in series, parallel, or combination, where each block represents the reliability of the component (=sensor) of the system at time t, $R_k(t)$.

This chapter illustrates how to incorporate availability based fault tolerance for heterogeneous sensor network collaboration to improve allocation system performance.

9.2 Availability Definition

9.2.1 Overview—Availability-Based Analysis

System availability and resilience have been studied extensively in the context of reliability engineering and manufacturing problems, but the fault tolerance approach for sensor network operation making it robust enough to loss of sensors and/or addition of new sensors is missing. In this book, reliability and availability models are used for heterogeneous sensor network collaboration to provide practical solutions for a fault tolerant allocation system performance.

In order to model TAPs' performance and ensure that it is robust to sensor failure, an availability-based analysis was conducted. The impact of each sensor on the system's availability was analyzed and optimized according to the results by reliability software optimization module [14, 16]. The reliability-based modeling and analysis approach was applied for the heterogeneous sensor collaboration for TAPs with three types of sensors (Fig. 9.2). The aim was to provide practical solutions for fault-tolerant system performance.

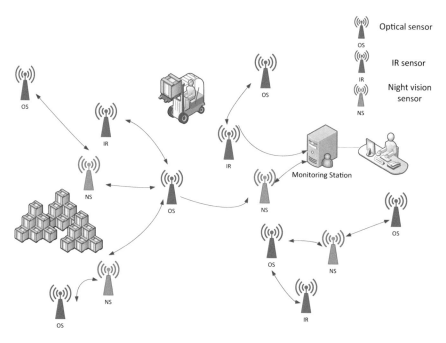

Fig. 9.2 Heterogeneous sensors supply network monitoring system with three different types of sensors: optical, infra-red, and night vision sensors (after Tkach et al. [14])

Fig. 9.3 RBD of a serial
system

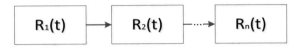

In this simulation, the aim is to analyze the performance of different algorithms aiming to develop an algorithm to improve the overall system availability to operate fault tolerantly up to 98% of the time.

To model the reliability structure of the system, a Reliability Block Diagram (RBD) is used. For a system to correctly function, every one of its components should properly operate, the respective system RBD should be represented as a series of its system components. Figure 9.3 illustrates a series structure of a system composed of n independent components, where $R_k(t)$ is the reliability function of each component [14].

A system with n components in series has reliability (at time t) equal to the product of the reliability of the components that compose it, as shown in Eq. (1):

$$R(t) = \prod_{k=1}^{n} R_k(t) \tag{9.1}$$

In a case where each component has an exponential failure distribution (which is the common way to represent the failure rate of electronic components), the reliability may be represented as:

$$R(t) = e^{-\lambda t} \tag{9.2}$$

where λ is the failure rate of the system, and is given by the fraction of sums of the failure rate of each system component:

$$\lambda(t) = \frac{1}{\sum_{k=1}^{n} \frac{1}{\lambda_k(t)}} \tag{9.3}$$

The system availability (4) is defined as the ratio between system uptime and downtime:

$$A = \frac{E[Uptime]}{E[Uptime] + E[Downtime]} = \frac{MTBF}{MTBF + MTTR} \tag{9.4}$$

where A is the system availability, *MTBF* is the mean time between failures, and *MTTR* is the mean time to repair. *MTBF* is inversely proportional to the failure rate (5):

$$MTBF = \frac{1}{\lambda(t)} \tag{9.5}$$

Table 9.1 MTBF and cost values summary

Type of sensor	MTBF	Cost	Number of sensors
Optical sensor	$\approx$28,000 h	$40	34
IR sensor	$\approx$163,000 h	$90	33
Night vision sensor	$\approx$212,000 h	$130	33

Consider the following MTBF values for three non-complementary sensors (optical, IR, and night vision) based on NPRD-2011 [9] and presented in Table 9.1. We derive the analytical solution of system availability which is given by:

$$MTBF = \frac{1}{\sum_{i=1}^{n} \frac{1}{MTBF_i}} = \frac{1}{34 \cdot \frac{1}{27862} + 33 \cdot \frac{1}{163132} + 33 \cdot \frac{1}{212765}} = 633.84 \quad (9.6)$$

$$A = \frac{E[Uptime]}{E[Uptime] + E[Downtime]} = \frac{MTBF}{MTBF + MTTR}$$
$$= \frac{633.84}{633.84 + (720 + 5)} = 0.46 \quad (9.7)$$

The system's availability value A is 46%, which means that 54% of the time the system will not be able to cover the entire area.

9.3 Monte Carlo Simulation

Monte Carlo simulation is a statistical analysis technique that was first used at Los Alamos National Laboratory in the 1940 s by scientists working on the development of an atomic bomb. It is widely used for optimization in engineering, finance, reliability and manufacturing among other fields [13]. Monte Carlo simulation refers to the Monte Carlo casino in Monaco [10]. The simulation process involves generating variables from a statistical distribution that exhibits random behaviors. Monte Carlo is also suitable for solving complex engineering problems because it can deal with a large number of random variables, various distribution types, and highly nonlinear engineering models.

Monte Carlo simulation performs a random sampling of possible values from the input probability distributions [11]. Each set of samples is called an iteration or history, and the resulting output from that sample is calculated. Monte Carlo simulation performs thousands of such histories, and the result is a probability distribution of possible outcomes that are available for statistical analysis.

9.3.1 Simulation Assumptions/Parameters

Simulation assumptions:

1. Failure and repair distributions of the components are exponential (corresponding to the typical failure of electronic devices). Hence, failure and repair rates are constant.
2. Upon failure, the sensor is repaired with normal time distributions of 240, 480, and 720 ± 2 h. This repair time includes transporting the failed sensor to a depot, its repair time, and transporting time back to the field.
3. Sensor replacement in the field takes 5 h (modeled as insertion distribution). This time duration was set in order to add the impact of a typical replacement operation on the system's availability.
4. Initial system deployment has no redundancies, in order to analyze the impact of adding redundant sensors to the system.

The experimental parameters used in this analysis are stated in Table 9.2. Task locations were randomly distributed with uniform distributions in the arena based on [12] and [2]. The sensor network was simulated in [15] for three sensor distributions: grid distribution uniformly distributed at random and normally distributed at random. Best performance was obtained for the uniform distribution; therefore, this distribution was considered in the current analysis. The sensor's operating distance values were set relative to the size of the arena (100 × 100 m). An inherent noise in each sensor was introduced according to typical operational distance ranges of known commercial sensors [17] relative to the arena size [14].

The simulation was performed for a period of 5 years of system operation (43,800 h) for a distributed sensor network with 100 sensors combined from three sensor types. Two types of sensor networks were considered to compare the influence of temporal restrictions on availability values.

1. Type 1—sensors are set to perform without temporal restrictions.

Table 9.2 Experimental parameters summary

Parameter	Values
Area dimensions	100 × 100 m
Total number of sensors	100
Types of sensors	Optical, IR, night vision
Sensors range coverage	Predefined from 15 to 45 m
Number of tasks	200
Simulation duration	200 steps
Tasks location	Uniformly distributed at random
Protocols	TRAP, ASAP, SRAP, STOP
Sensor's location	Uniformly distributed random
Tasks arrival time	Every 1 step

2. Type 2—sensors with low MTBF values (optical sensors) are set aside and come into performance only when needed.

9.4 Results and Discussion

Independent Monte Carlo runs with 100 repetitions were executed. Availability analysis was conducted using a Monte Carlo simulation in 1-min intervals. Results are calculated with the statistical significance of 95%.

Results (Figs. 9.4 and 9.5) indicated that the average availability increased from 38 to 86% when low MTBF sensors came into action only when necessary. The system average availability was increased to 98% when 6 spare sensors were used. As both systems start with the same initial values, they start from 100% availability. Sensors in a system of type 1 operate based on a regular algorithm and therefore sensors with low MTBF values start to malfunction and decrease system availability much faster than in a system of type 2. In Fig. 9.4, after a period of time, the availability starts to increase as a result of sensors that have been repaired. Also, in a system of type 1,

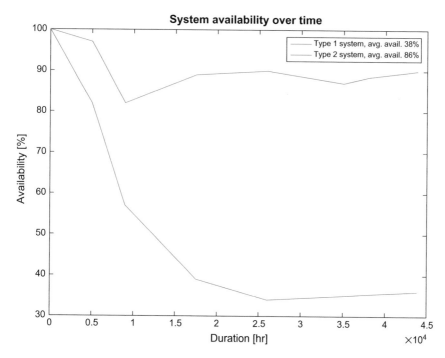

Fig. 9.4 System's average availability over time based on different network types—same rest time for all sensors and sensors with low MTBF values operate only when necessary. The dots represent the time interval at which sensors were repaired (after Tkach et al. [14])

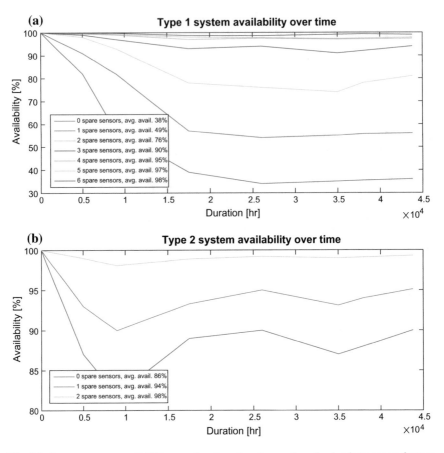

Fig. 9.5 System average availability over time based on the number of redundant sensors for two system types. **a** type 1 system; **b** type 2 system. The dots represent the time interval at which sensors were repaired (after Tkach et al. [14])

the failure rate of sensors is higher than the repair rate, so its availability is almost constantly decreased.

A redundancy of sensors is needed to improve system availability. The influence of adding redundant sensors on the availability of the two system types was analyzed and compared (Fig. 9.5).

Figure 9.5 illustrates the sensitivity analysis of the influence of the redundant number of sensors on the two types of sensor networks. The availability of both systems increases with the increase in the number of redundant sensors until a high availability value is reached. The initial decrease of system availability is due to the malfunction of the sensors in the system; after these sensors are repaired, the availability starts to increase. Different repair times of 240, 480, and 720 h. [3] have been considered for sensitivity analysis. These repair times include travel time, transporting of the failed sensor to a depot, lead time, repair time, and transporting

Table 9.3 The influence of the number of redundant optical sensors on type 1 systems' availability with different repair times (240, 480, and 720 h)

Quantity of redundant sensors	Availability as a function of repair time		
	240 h (%)	480 h (%)	720 h (%)
0	72	51	38
1	89	64	49
2	96	82	76
3	98	93	90
4	99	96	95
5	99	98	97
6	99	99	98

Table 9.4 Optimization results for systems' availability of 98% with 720 h repair time

Type of system	Quantity of redundant sensors
Type 1 system	6 optical sensors
Type 2 system	2 optical sensors

time back to the field. As in the system of type 1, the sensors malfunction more frequently, and initial availability decreases for a longer time duration. Tables 9.3 and 9.4 summarize the redundancy recommendation for the system and the availability achieved:

Based on the results, in order to achieve system availability above 98% over a 5-year period, the systems should include 106 and 102 sensors, respectively, for type 1 and type 2 systems.

After redundant sensors are introduced, the system becomes parallel. The respective system RBD is updated according to the new system components. Figure 9.6 illustrates the updated structure of a system composed of n independent components, where $R_k(t)$ is the reliability function of each component. The new product of the reliability of the components that compose the system is shown in Eq. (8):

$$R(t) = \prod_{k=1}^{n-m} R_k(t) \times \left\{ 1 - \prod_{p=1}^{m} [1 - R_p(t)] \right\} \tag{9.8}$$

Fig. 9.6 RBD of the parallel system

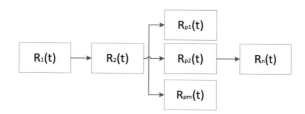

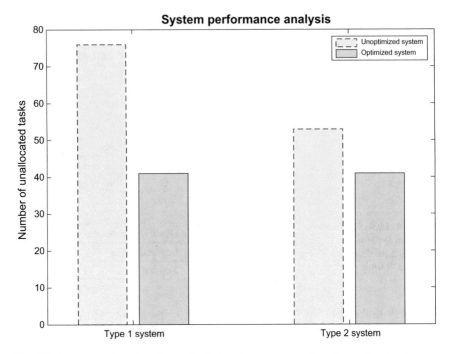

Fig. 9.7 Average number of unallocated tasks based on system configuration and system type

Results of the dual-layer system applying the new approach indicate that for both system types the optimized systems processed on average more tasks than the corresponding un-optimized systems (Fig. 9.7). Optimized type 1 systems resulted in 165.4% more processed tasks, and type 2 systems resulted in 13.1% more processed tasks when availability analysis results were implemented to the system (F = 9.34, $p = 0.02$). These findings indicate that the fault tolerance of the multi-sensor system was improved, contributing to system performance improvement.

References

1. Beaudet S, Courtney T, Sanders W (2006) A behavior-based process for evaluating availability achievement risk using stochastic activity networks. In: Proceedings of RAMS, Newport Beach, California, pp 23–26
2. Byers J, Nasser G (2000) Utility-based decision-making in wireless sensor networks. In: Proceedings of the 1st ACM international symposium on mobile ad hoc networking & computing. IEEE Press, Boston, MA, USA, pp 143–144
3. Cadwallader LC (2012) Review of maintenance and repair times for components in technological facilities. INL/EXT-12–27734, Idaho National Laboratory
4. Dâmaso A, Rosa N, Maciel P (2014) Reliability of wireless sensor networks. Sensors 14(9):15760–15785

5. Dias MB, Zinck M, Zlot R, Stentz A (2004) Robust multirobot coordination in dynamic environments. In: IEEE proceedings of international conference on robotics and automation, pp 3435–3442
6. Edan Y, Nof SY (2000) Sensor economy principles and selection procedures. IIE Trans Des Eng 32(3):195–203
7. Jeong W, Nof SY (2009) A collaborative sensor network middleware for automated production systems. Int J Comput Ind Eng 57:106–113
8. Levalle RR (2018) Resilience by teaming framework. Resilience by teaming in supply chains and networks. Springer, Cham, pp 59–64
9. Mahar D, Fields W, Reade J, Zarubin P, McCombie S (2011) Nonelectronic parts reliability data. Reliability Information Analysis Center
10. Metropolis N (1987) The beginning of the Monte Carlo method. Los Alamos Sci 125–130
11. Morin RL (2019) Monte Carlo simulation in the radiological sciences. CRC Press
12. Rowaihy H, Eswaran S, Johnson M, Verma D, Bar-Noy A, Brown T, Porta TL (2007) A survey of sensor selection schemes in wireless sensor networks. Proc SPIE 6562
13. Sobol IM (2018) A primer for the Monte Carlo method. CRC Press
14. Tkach I, Edan Y, Nof SY (2017) Multi-sensor task allocation framework for supply networks security using task administration protocols. Int J Prod Res 55:5202–5224
15. Tkach I, Jevtić A, Edan Y, Nof SY (2018) A modified distributed bees algorithm for multi-sensor task allocation. Sensors 18(3):759. https://doi.org/10.3390/s18030759
16. Tkach I, Nof SY, Edan Y (2013) Fault tolerant task allocation in a multisensory supply chain monitoring system. In: International Conference of Production Research, Iguacu, Brazil
17. Yoon Y, Kim YH (2013) An efficient genetic algorithm for maximum coverage deployment in wireless sensor networks. IEEE Trans Cybern 43(5):1473–1483
18. Zhang S, Liu Y, Pu J, Zeng X, Xiong Z (2009) An enhanced coverage control protocol for wireless sensor networks. In: IEEE International Conference on System Sciences, pp 1–7

Chapter 10
Analytical Analysis of a Simplified Scenario of Two Sensors and Two Tasks

This chapter describes and analyzes analytically the performance of HDBA in 'Layer 1' and the performances of TAPs in 'Layer 2' of the task allocation system (Fig. 10.1). An analysis of a simplified system consisting of two sensors and two tasks is presented. This analysis was conducted to reveal the basic operating mechanism of HDBA and to analyze the benefit of the collaboration of agents in the allocation

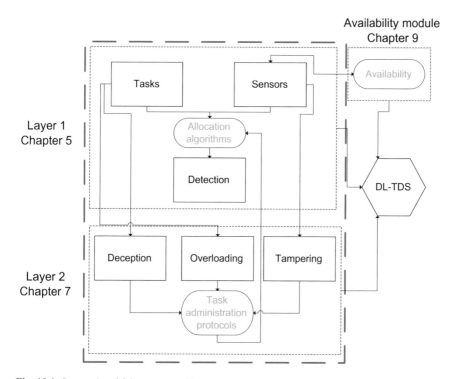

Fig. 10.1 Layers 1 and 2 in system architecture

© Springer Nature Switzerland AG 2020
I. Tkach and Y. Edan, *Distributed Heterogeneous Multi Sensor Task Allocation Systems*,
Automation, Collaboration, & E-Services 7,
https://doi.org/10.1007/978-3-030-34735-2_10

process in 'Layer 1' and the performance of a monitoring layer that is dedicated to handling special cases by applying TAPs in 'Layer 2'.

10.1   Single-Layer Performance Analysis

Analytical analysis of a small scenario that includes two sensors and two tasks is presented. Consider the example presented in Fig. 10.2 that includes an arena with two sensors with different performances, and two tasks with different priorities and distances from the sensors. For simplicity of the example, we set the initial completion time value t_{init} of the tasks to 1.

The normalized priority of task 1 is F_1 and task 2 is F_2. The Euclidian distance of sensor 1 from task 1 and task 2 is D_{11} and D_{21}, respectively. The Euclidian distance of sensor 2 from task 1 and task 2 is D_{12} and D_{22}, respectively. The performance of sensor 1 on task 1 and task 2 is V_{11} and V_{21}, respectively. The performance of sensor 2 on task 1 and task 2 is V_{12} and V_{22}, respectively.

Since the penalty for synchronization and setup times are strongly influenced by the hardware and the environment, they can vary significantly in different systems. Due to this fact, they were disregarded in this simplified analysis that represents a general case of two sensors and two tasks.

The derived probabilities of the sensors to be allocated to tasks in HDBA are (for simplicity, we assume that $\alpha, \beta, \gamma = 1$):

The probability of sensor 1 to be allocated to task 1 is defined by

$$p_{11} = \frac{F_1\left(\frac{1}{D_{11}}\right)V_{11}}{F_1\left(\frac{1}{D_{11}}\right)V_{11} + F_2\left(\frac{1}{D_{21}}\right)V_{21}} \tag{10.1}$$

The probability of sensor 1 to be allocated to task 2 is defined by

Fig. 10.2 Two sensors and
two tasks problem

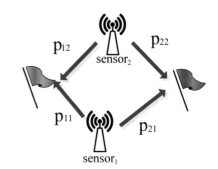

$$p_{21} = \frac{F_2\left(\frac{1}{D_{21}}\right)V_{21}}{F_1\left(\frac{1}{D_{11}}\right)V_{11} + F_2\left(\frac{1}{D_{21}}\right)V_{21}}$$ (10.2)

The probability of sensor 2 to be allocated to task 1 is defined by

$$p_{12} = \frac{F_1\left(\frac{1}{D_{12}}\right)V_{12}}{F_1\left(\frac{1}{D_{12}}\right)V_{12} + F_2\left(\frac{1}{D_{22}}\right)V_{22}}$$ (10.3)

The probability of sensor 2 to be allocated to task 2 is defined by:

$$p_{22} = \frac{F_2\left(\frac{1}{D_{22}}\right)V_{22}}{F_1\left(\frac{1}{D_{12}}\right)V_{12} + F_2\left(\frac{1}{D_{22}}\right)V_{22}}$$ (10.4)

If we disregard the wheel-selection rule, the completion times of the tasks are derived for 4 following cases:

1. $p_{11} > p_{21}$ and $p_{12} > p_{22}$

2. $p_{11} < p_{21}$ and $p_{12} < p_{22}$

3. $p_{11} > p_{21}$ and $p_{12} < p_{22}$

4. $p_{11} < p_{21}$ and $p_{12} > p_{22}$

The solution for each case is given by:

1. At first, sensors 1 and 2 will be allocated to task 1, therefore the task completion times t_c and the overall completion time T are given as:

$$t_{c1} = \frac{1}{V_{11} + V_{12}}; t_{c2} = \frac{1}{V_{11} + V_{12}} + \frac{1}{V_{21} + V_{22}}$$

$$T = \frac{2}{V_{11} + V_{12}} + \frac{1}{V_{21} + V_{22}}$$

2. At first, sensors 1 and 2 will be allocated to task 2, therefore the task completion times t_c and the overall completion time T are given as:

$$t_{c1} = \frac{1}{V_{11} + V_{12}} + \frac{1}{V_{21} + V_{22}}; t_{c2} = \frac{1}{V_{21} + V_{22}}$$

$$T = \frac{1}{V_{11} + V_{12}} + \frac{2}{V_{21} + V_{22}}$$

3. At first, sensors 1 and 2 will be allocated to task 1 and task 2, respectively. Therefore, the task completion times t_c and the overall completion time T are given as:

   a. if the sensor performance values derive $V_{11} < V_{22}$ then sensor 2 will complete task 2 before sensor 1 completes task 1, therefore:

$$t_{c1} = \frac{1}{V_{11} + V_{12}} + \frac{1}{V_{22}}; t_{c2} = \frac{1}{V_{22}}$$

$$T = \frac{1}{V_{11} + V_{12}} + \frac{2}{V_{22}}$$

   b. if the sensor performance values derive $V_{11} > V_{22}$ then sensor 1 will complete task 1 before sensor 2 completes task 2, therefore:

$$t_{c1} = \frac{1}{V_{11}}; t_{c2} = \frac{1}{V_{11}} + \frac{1}{V_{21} + V_{22}}$$

$$T = \frac{2}{V_{11}} + \frac{1}{V_{21} + V_{22}}$$

   c. if the sensor performance values derive $V_{11} = V_{22}$ then sensor 1 will complete task 1 at the same time sensor 2 will complete task 2, therefore:

$$t_{c1} = \frac{1}{V_{11}}; t_{c2} = \frac{1}{V_{22}}$$

$$T = \frac{1}{V_{11}} + \frac{1}{V_{22}}$$

4. a. if the sensor performance values derive $V_{12} < V_{21}$ then sensor 1 will complete task 2 before sensor 2 completes task 1, therefore:

$$t_{c1} = \frac{1}{V_{11} + V_{12}} + \frac{1}{V_{21}}; t_{c2} = \frac{1}{V_{21}}$$

$$T = \frac{1}{V_{11} + V_{12}} + \frac{2}{V_{21}}$$

   b. if the sensor performance values derive $V_{12} > V_{21}$ then sensor 2 will complete task 1 before sensor 1 completes task 2, therefore:

$$t_{c1} = \frac{1}{V_{12}}; t_{c2} = \frac{1}{V_{12}} + \frac{1}{V_{21} + V_{22}}$$

$$T = \frac{2}{V_{12}} + \frac{1}{V_{21} + V_{22}}$$

c. if the sensor performance values derive $V_{12} = V_{21}$ then sensor 1 will complete task 2 at the same time sensor 2 will complete task 1, therefore:

$$t_{c1} = \frac{1}{V_{12}}; \, t_{c2} = \frac{1}{V_{21}}$$

$$T = \frac{1}{V_{12}} + \frac{1}{V_{21}}$$

Considering these cases, a collaboration of sensors to complete tasks while both sensors are allocated to the first task and afterwards to the second task will yield better results except in cases where $V_{11} = V_{22}$ and $V_{12} = V_{21}$. The overall completion times in cases of collaboration are shorter than in cases without it, as can be seen in the following inequalities:

$$\frac{2}{V_{11} + V_{12}} + \frac{1}{V_{21} + V_{22}} < \frac{2}{V_{11}} + \frac{1}{V_{21} + V_{22}} \tag{10.5}$$

$$\frac{1}{V_{11} + V_{12}} + \frac{2}{V_{21} + V_{22}} < \frac{1}{V_{11} + V_{12}} + \frac{2}{V_{21}} \tag{10.6}$$

$$\frac{2}{V_{11} + V_{12}} + \frac{1}{V_{21} + V_{22}} < \frac{2}{V_{12}} + \frac{1}{V_{21} + V_{22}} \tag{10.7}$$

$$\frac{1}{V_{11} + V_{12}} + \frac{2}{V_{21} + V_{22}} < \frac{1}{V_{11} + V_{12}} + \frac{2}{V_{22}} \tag{10.8}$$

A quantitative example of this collaboration is as follows:

Considering example case #4 and given the performance values $V_{12} = V_{11} = V_{22} = 1$ and $V_{21} = 2$, full collaboration $T = \frac{2}{V_{11} + V_{12}} + \frac{1}{V_{21} + V_{22}}$ and $T = \frac{1}{V_{11} + V_{12}} + \frac{2}{V_{21} + V_{22}}$ result in overall completion times of $T = 1.33$ and $T = 1.16$, respectively, while in the partial collaboration or non-collaboration case $T = \frac{1}{V_{11} + V_{12}} + \frac{2}{V_{21}}$, $T = \frac{2}{V_{12}} + \frac{1}{V_{21} + V_{22}}$ and $T = \frac{1}{V_{12}} + \frac{1}{V_{21}}$ result in $T = 1.5$.

The example above, while very simple, indicates that there are cases where cooperation between sensors is beneficial. This is true especially in cases where multiple tasks need to be handled concurrently.

In contrast to a greedy algorithm, the inherent statistical property of HDBA, which is using the wheel-selection rule, enables it to search the solution space and to find a near-optimal solution that considers cooperation.

10.2   Dual-Layer Performance Analysis

A dual-layer heterogeneous multi-sensor framework includes a process layer that executes HDBA and a monitoring layer that is dedicated to handling special cases of problems in sensors and in the allocation process by applying TAPs. The special cases are defined in this book as:

(1) Overloading.
(2) Deception.
(3) Tampering.

To analyze the benefits of applying TAPs, these three special cases were applied to the simplified scenario:

1. Overloading
   If $p_{11} > p_{21}$ and $p_{12} > p_{22}$ and task 1 is overloaded then both sensors are allocated to task 1 and are doomed to this allocation without the ability to complete it. The completion times of the tasks for this case are:

$$t_{c1} = \infty; t_{c2} = \infty$$

By applying STOP, the allocation of both sensors will be interrupted by the protocol, and the new task completion times are:

$$t_{c1} = \mu_1 + 2\sigma_1; t_{c2} = \mu_1 + 2\sigma_1 + \frac{1}{V_{21} + V_{22}}$$

2. Deception
   If $p_{11} > p_{21}$ and $p_{12} > p_{22}$ and task 2 needs to be handled urgently, then both sensors will be allocated to task 1 until its completion, despite the urgency of task 2, due to the inherent properties of the allocation algorithms. The task completion times for this case are:

$$t_{c1} = \frac{1}{V_{11} + V_{12}}; t_{c2} = \frac{1}{V_{11} + V_{12}} + \frac{1}{V_{21} + V_{22}}$$

By applying TRAP, the urgency of task 2 will be considered and the allocation of both sensors will be directed to it. The new task completion times for this case are:

$$t_{c1} = \frac{1}{V_{11} + V_{12}} + \frac{1}{V_{21} + V_{22}}; t_{c2} = \frac{1}{V_{21} + V_{22}}$$

It is evident that task 2 has been completed more quickly.

3. Tampering
   If $p_{11} > p_{21}$ and $p_{12} > p_{22}$ and sensor 1 has failed, then both tasks will be completed by sensor 2 solely. The task completion times for this case are:

$$t_{c1} = \frac{1}{V_{12}}; t_{c2} = \frac{1}{V_{12}} + \frac{1}{V_{22}}$$

By applying ASAP, a back-up sensor will come into action. The new task completion times for this case are:

$$t_{c1} = \frac{1}{V_{11} + V_{12}}; t_{c2} = \frac{1}{V_{11} + V_{12}} + \frac{1}{V_{21} + V_{22}}$$

It is evident that both tasks have been completed more quickly.

This simple analytical analysis of the dual-layer indicates that applying TAPs to special cases can considerably improve systems' performance.

Chapter 11
An Outlook of Multi-sensor Task Allocation

There are plentiful applications of multi-sensor task allocation in practice. The previous chapters of this book specified, formulated, described and illustrated the impacts of optimal multi-sensor task allocation on the quality and performance of supply networks and law enforcement systems. Emerging applications and technologies of the future will include sensors with different abilities that need to be dynamically allocated to various and multiple tasks. Some of these emerging technologies include: collaborative telepresence combining Augmented Reality (AR), Virtual Reality (VR), 5G networks and advanced sensors to enable remote work, remote healthcare and remote communication; advanced food tracking by sensors to monitor every step of a food item's progress through the supply network; advanced packaging and order monitoring of the quality of goods via sensors; safety monitoring in nuclear and chemical reactors by distributed sensors to detect abnormal activities; smart cities will be enabled by sensors exploitation for safety, efficiency, traffic and lighting control; water and wastewater monitoring; fire detection; smart logistics; smart manufacturing and Industry 4.0; robotic navigation; conflict and error prevention; autonomous driving with sensor embedded vehicles and future combat systems. These technologies will be significantly influenced by the efficient allocation of sensors and tasks and will benefit by applying the dual-layer system architecture as described in this book.

11.1 Summary

A dual-layer system framework aimed at providing high quality allocations of sensors for task allocation that includes a process layer that allocates sensors to tasks and a monitoring layer that is dedicated to handling special cases has been described, implemented, and analyzed extensively. The framework enables allocating tasks in all locations of the area of interest independent of the sensor network distribution. A collection of activities to control such system performance is discussed and a

© Springer Nature Switzerland AG 2020

I. Tkach and Y. Edan, *Distributed Heterogeneous Multi Sensor Task Allocation Systems*,
Automation, Collaboration, & E-Services 7,
https://doi.org/10.1007/978-3-030-34735-2_11

system objective function is presented. The correct way to implement task administration protocols to optimize the objective function within the framework is presented. "TRAP" manages task priorities to deal with the deception of sensors and allows better allocation of sensors to the most important tasks. "ASAP" ensures that tasks will be treated as soon as possible and will not be unnecessarily delayed due to tampering of sensors. "SRAP" allocates the best match of sensors to handle tasks based on the HDBA algorithm, and "STOP" was designed to deal with attempts of overloading of agents, and ensuring optimal sensors availability in the system by time-out policy. The framework enables dynamic task management and synchronization to optimally allocate and manage the a priori distributed sensors to minimize handling times and maximize the number of correctly allocated tasks.

It was shown that the HDBA algorithm efficiently allocated a large group of heterogeneous sensors to upcoming tasks, which are unknown in their spatial and temporal distributions; it provides scalability in terms of the number of tasks and sensors. It resulted in a statistically significant 6.6% better performance in terms of task completion time using 100 sensors in a uniform distribution, with respect to the second-best algorithm, and 18.3% better with respect to the Greedy algorithm, which performed the worst. The control parameters, α, β, and γ provide a mechanism to adjust the sensor swarm behavior and bias importance of the priority, distance, and sensors' performance, respectively. Simulation results revealed that the best task completion time was achieved by using $\alpha = \beta = \gamma = 1$. Uniform distribution of the sensors resulted in better system performance with the shortest task completion time, with 5.1 and 26.3% improvement from the grid and normal distributions, respectively ($p < 0.05$).

HDBA was tested for two TSP benchmark instances: Berlin52 with 52 locations and A280 with 280 locations, using TSPLIB as a reference. HDBA resulted in near-optimal results in handling this NP-hard optimization problem and provided solutions for the two instances. HDBA managed to find solutions without any prior knowledge of the solution space or the help of any local search routine. Through the instances, the algorithm maintained a steadily improved behavior, and it resulted as a second-best algorithm for the large instance of A280 after BS, indicating its fitness for solving TSP optimization efficiently.

HDBA was applied for solving the LEP problem inspired by real police logs by allocating heterogeneous police officers to dynamic tasks whose locations, arrival times, and importance levels are unknown a priori. Realistic measurements were obtained by Amador and Zivan [1] by consulting police officers about specific evaluation metrics that are of interest to them. Three algorithms for solving the LEP problem were compared according to five metrics: team utility, an average execution delay, percentage of abandoned tasks, percentage of shared tasks, and the average arrival time of agents to tasks. The application was shown to be effective in allocating dynamic tasks to heterogeneous police agents using simulation analyses. HDBA generated more collaboration among agents and resulted in better performance in terms of team utility and execution delay than both FMC_TA^{H+} and SA, with statistically significant 6 and 41% better team utility in the highest shift load compared to FMC_TA^{H+} and SA, respectively.

The dual-layer system analyses revealed that TAPs increase the system performance in the scenarios of deception, tampering, and overloading by more than 80% of the number of successful treatments for the most important tasks and decreases by more than 72% the number of unallocated tasks in comparison to a single-layer system.

Reliability analysis performed using Monte Carlo simulation for the heterogeneous sensor network indicates that overall system availability was statistically significantly improved, ensuring fault tolerant system operation. Simulation results of TAPs operation indicated a statistically significant increased number of handled tasks (by 13.1%, $p = 0.02$) when reliability analysis recommendations were applied.

Table 11.1 presents the relation between problems in multi-sensor task allocation, book chapters, and applications.

Table 11.1 The relation between problems in multi-sensor task allocation and book chapters and applications

Problem	Concepts and methods	Application
An optimal way to control a heterogeneous sensor network to work adaptively in real time without knowing a priori when, where, and which task will enter or leave the environment and deal with the difficulties of collaboration that arise when simultaneously treating several tasks in a crowded dynamic environment	5. Single-layer multi-sensor task allocation system 5.3 Algorithms for multi-sensor task allocation	6.2 Multi-sensor task allocation 6.3 Traveling salesman problem 6.4 Law enforcement problem
Disturbances in sensors and in the allocation process such as very long attendance times and many tasks with the same priorities	7. Dual-layer multi-sensor task allocation system 7.3 Task Administration protocols	8. Multi-sensor task allocation in the supply network Deception Tampering Overloading
Fault tolerant sensor operation to make it robust enough to loss of sensors and/or addition of new sensors	9. Fault tolerant multi-sensor system with high availability 9.3 Availability	9.4 Multi-sensor task allocation in the supply network Type 1 system Type 2 system

11.2   Final Remarks

The system developed in this book provides several advantages. It provides the ability to allocate large numbers/swarms of heterogeneous sensors to tasks in real time when information is not known a priori. It allows dynamic reallocation of sensors to tasks and to handle new tasks with quick response and adaptation to dynamic conditions. It allows accommodation of addition/subtraction of sensors during operation and allocation of a limited number of sensors with limited range coverage to treat dynamic tasks. The system was shown to be robust to sensor failures and able to deal with cases of priority reassignment and time-outing of tasks in real time that traditional allocation algorithms are incapable of.

The analytical analysis revealed that there are cases where cooperation between sensors is beneficial, especially in cases where multiple tasks need to be handled concurrently and that applying TAPs to special cases can considerably improve system performance.

The limitation of the proposed HDBA algorithm is its probabilistic nature. Due to its decentralized nature, HDBA cannot guarantee an optimal solution, but it can provide sub-optimal solutions to NP-hard problems; it yielded better results in comparison to other algorithms analyzed in this book. The decentralized decision-making process of the agents ensures system robustness and reliability in case of single sensor failure. Another limitation is that the communication regarding the estimated location and priority of the tasks is done by broadcasting, which is a centralized form of communication.

The logic and control parameters of TAPs are designed to fit the specific case studies used in this book. The structure of these protocols can be used for different scenarios, but the parameters should be adjusted for a specific case study and be optimized based on the objective of the specific system for optimal results.

With this new framework, system designers can adjust and test the system for desired or required availability, yielding a quantitative tool for controlling the number of processed tasks and controlling the system's fault tolerance level. It enables designers to determine how many sensors are necessary for a decentralized system to achieve desired system performance. This framework is essential for systems that require continuous operation, e.g., monitoring systems, lifesaving systems, complex systems with different types of agents and complex formations, and systems that require high availability or demand for specific fault tolerance values. In case very low-cost agents are applied, where quantity is not an issue, and when low performance and availability are acceptable this framework would not be essential.

The main contributions of DL-TDS are the development of:

1. An adaptable framework, applicable to physical and digital systems that compute the expected value of system performance given the sensor, environmental, and task parameters.
2. A swarm intelligence heterogeneous agent algorithm to decide in real time which sensor addresses which task without knowing a priori when, where, and which task will enter or leave the environment.

3. A suite of task administration protocols to handle risks and problems within the sensory system such as sensor availability, conflicts in tasks' priorities, and high time-consuming tasks.
4. A dual-layer system to ensure fault tolerant sensor operation that is robust enough to sensor failure, deception, overloading, and tampering.

11.3 Future Research Directions

Multi-sensor systems and task allocation is a rapidly growing area. Future multi-sensor systems will become more complex with a high number of low-cost sensors and the need for coordination and allocation will grow. Some research directions can be implemented and expanded in the future:

1. Implement the multi-sensor task allocation framework on an operational monitoring system based on a swarm of robots to allocate upcoming tasks or tasks with unknown a priori locations, priorities, and arrival times for the robots. The robots should possess proper hardware to be able to execute basic decisions regarding the allocation. The formation must be distributed and tested for various deployment structures of robots. The robots must have sensors of different types to be able to recognize tasks.

   The experimental setup should include allocation in different environmental conditions, i.e., daylight, night time, heat inflating tasks, etc.

   All should be tested for the performance measures of tasks completion time, number of unallocated tasks, number of tasks allocated to sensors and compared to the simulation results of this book.

   This experiment can include scalability evaluation, which can be achieved by physically removing several robots during the experiment, and adding new robots at different times.

   Additional evaluations can include problems within the system of overloading, deception, and tampering. These should be introduced as high time-consuming tasks, dynamic change in priorities of some tasks during execution, and failure of some robots.

   The control parameters of the HDBA should be adjusted during the experiment in order to calibrate the performance of the robotic swarm to the actual experimental settings.
2. Analyze the effect of communication delay between sensors, which can increase task completion time.

   A communication delay parameter can be introduced to the simulation of the system. The value of this communication delay should be derived from the current hardware limitations. This parameter should be added to the allocation time value of the system, and can affect the performance by increasing the task completion times:

$$T = \sum_{i=1}^{M} t_{c_i} + t_{d_i}$$

where t_{di} is the communication delay of sensors involved in the allocation.

3. Analyze the effect of different stopping criteria of the HDBA on the task completion time. Different stopping criteria can result in different termination times and therefore affect the completion times.

   An experimental setup of different stopping criteria should be built. It should include different numbers of simulation cycles allowed for HDBA execution. Each stopping criterion should be evaluated for 100 independent simulation runs, and the mean performance value should be presented. Then, a sensitivity analysis of the stopping criteria should be conducted, analyzing the effect of increasing the allowed number of simulation cycles on the system performance for performance measures indicated above.

4. Apply different sensor topologies and analyze their influence on allocation performance.

   Different topologies can affect situation awareness, latency, and communication, and therefore can have an impact on the performance of the system.

   Different topologies of sensors can be introduced and evaluated. These can include topologies with different numbers of sensors and different communication routing or broadcasting schemes. Examples of such topologies include star topology, cluster-based topology, tree-based topology, to name a few. Each topology has its own communication pattern and can affect the performance in terms of time delay and sensory awareness of the environment and tasks. Each topology should be implemented to the sensory system and simulated for the performance measures indicated above.

5. Finding the threshold (in time units/cycle numbers) in HDBA required for minimal accepted allocation performance.

   First, the minimal accepted allocation performance of the system can be evaluated. Then, sensitivity analysis of a threshold can be conducted based on the time units and the number of simulation cycles allowed for the algorithm to run. The threshold may be affected by the hardware and the software used for simulation; therefore these values are platform dependent.

6. Design of autonomous learning systems to dynamically adjust control parameters for HDBA and TAPs, and anticipate possible changes in the system structure/function. This minimizes human intervention and system calibration and enables smart scenario-analysis to fine-tune control protocols, even in the face of system changes.

   This can be done by experiments for investigating the effect of the control parameters on system performance for different system structures and different goals (minimizing allocation times, maximizing the allocation of most important tasks, minimizing failures of best performing sensors, etc.).

7. A penalty can be introduced for reallocating a sensor from a task during task execution. This can influence the execution times and the allocation strategy, since sensors will be constrained in their reallocation.
8. Optimization of HDBA parameters for solving LEP. This must include the optimization of the control parameters α, β, and γ that provide a mechanism to adjust the swarm behavior and bias the importance of the urgency, distance, and skills, respectively. The optimization of these parameters can result in the reduction of the execution delay of important tasks of types 1 and 2 in LEP, improving the overall utility.
9. Adding a second layer of TAPs to LEP to improve performance and handle problems in the allocation process.

Reference

1. Amador S, Zivan R (2017). Incentivizing cooperation between heterogeneous agents in dynamic task allocation. In: Proceedings of 16th international conference on autonomous agents and multiagent systems (AAMAS), São Paulo, Brazil, pp 1082–1090

Appendix
Notation

Acronyms

ACO	Ant Colony Optimization
ASAP	Assignment Analysis Protocol
BS	Bees System
CI	Confidence interval
DBA	Distributed Bees Algorithm
DCP	Deception
DL-TDS	Dual-layer task allocation system
FLIR	Forward Looking Infrared Radar
FMC_TA^{H+}	Fisher Market Clearing for Task Allocation with Heterogeneous Agent
GA	Genetic Algorithm
HDBA	Heterogeneous Distributed Bees Algorithm
LEP	Law Enforcement Problem with Heterogeneous Agents
ML	Monitoring Layer
MTBF	Mean Time Between Failures
MTTR	Mean time to repair
OVL	Overloading
PL	Process Layer
RBD	Reliability Block Diagram
SA	Simulated Annealing
SRAP	Sensors Allocation Protocol
STOP	Synchronization and Time-Out Protocol
TAPs	Task Administration Protocols
TMP	Tampering
TRAP	Task requirement Protocol
TSP	Traveling salesman problem

© Springer Nature Switzerland AG 2020
I. Tkach and Y. Edan, *Distributed Heterogeneous Multi Sensor Task Allocation Systems*,
Automation, Collaboration, & E-Services 7,
https://doi.org/10.1007/978-3-030-34735-2

Variables

A	Availability
Cap	Capability function
CT_k	Current task being performed by agent k
D_{ik}	Euclidean distance between the kth sensor and the ith task
F_i	Normalized priority of task i in the queue
f_i	Priority of task i
H	Sum of the priorities of tasks that were successfully completed
$I(\Psi)$	Importance of task
i	Index of tasks
k	Index of sensors/agents
L_k	Tour length of the kth ant
M	Set of tasks
N	Set of sensors
p_i^k	Probability of agent k to be allocated to task i
Q_s	Assignment analysis function
$\vec{q}$	Number of agents performing a task
R	Allocation range
$R_k(t)$	Reliability at time t
S	Collective performance of the sensors allocated to tasks
s	Skill of agent
T	Sum of tasks completion time
t	Temperature in SA
t_{c_i}	Task i completion time
tct_i	Required completion time for task i
$t_{current}$	Current time
t_{ex_i}	Remaining execution time of task i
$t_{i_{arrival}}$	Arrival time of task i
$t_{i_{f\_completion}}$	Full completion time of task i
t_{init_i}	Initial time value of task i required for its allocation
to	Time-out value
t_{p_i}	Elapsed time of task i execution
tst	Task start time
u_{ik}	Fisher Market utility function
V_I	Collective performance of the system
$V_{I S_{default}}$	Default allocation performance
V_{ik}	kth sensor's performance on the ith task
V_{os}	Optimal set of sensors performance
V_t	Waiting time penalty value
W_j	Workload set
w^s	Workload of specific skill
x_{ik}	Amount of task that i is allocated to agent k
α	Parameter that control the relative importance of priority

β	Parameter that control the relative importance of distance
Γ_i	Control parameter in Fisher Market
γ	Parameter that control the relative importance of sensor performance
Δ_i	Deadline value of task i
$\Delta \upsilon_i^k$	Quantity per unit of trail laid on sensor i by the kth ant
Δw	Amount of completed work
δ	Control parameter in Market based algorithm
ε	Number of unallocated tasks
η	Assignment type request
θ	Threshold value
$\lambda(t)$	Failure rate of the system
μ_{ct}	Task mean completion time
Ξ	Constant
ξ_i	Visibility of ith ant
π	Penalty for task interruption
ρ	Evaporation of trail between time t and $t + n$
σ_{ct}	Standard deviation of task completion time
σ^k	Schedule of agent
τ	Waiting time
υ_i	Intensity of ant's i trail at time t
υ_o	Intensity threshold
Φ	Price in Fisher Market
φ	Bias parameter for the importance of H relative to S
χ_k	Number of treated tasks by the kth sensor
Ψ_i	Task
ψ	Parameter that control the relative importance of trail
ω	Parameter that control the relative importance of visibility
$\mathcal{R}$	Real numbers

Index

A

Acoustic sensor, 53, 95

Agent, xi, xii, 1, 5, 9–11, 20, 21, 27–34, 36, 39, 40, 63, 68–71, 82, 84, 128

Allocation, ix–xii, 1–5, 9–11, 15–17, 19–21, 24–27, 29, 32–36, 38, 39, 41, 49, 52–54, 56, 57, 60, 68–72, 76, 82, 84, 85, 87–89, 93, 95, 105–107, 117, 122, 125, 127–131, 133, 134

Ant colony optimization, 33, 40

Artificial intelligence, 29

Assignment, 1, 2, 51, 68, 87–89

Augmented reality, 125

Availability, x, xi, 3, 4, 6, 16–22, 25, 56, 84, 87, 88, 105–114, 127–129, 134

B

Bees algorithm, 10, 30, 31, 33, 133

Bee system, 33, 39, 53

Benchmark, x, 20, 21, 61, 71, 72, 76, 93, 126

Bio-inspired approaches, 10

Broadcast communication, 18, 25, 26, 28, 31, 34

C

Centralized, 9, 11

Collaboration, vii, 9, 29, 30, 51, 56, 69, 76, 87, 88, 106, 107, 117, 121, 126, 127

Collective behavior, 10, 30

Communication, 11, 26, 30–34, 51, 68, 125, 128–130

Complex systems, 1, 106, 128

Contract net protocol, 11

Control parameters, 33, 35, 39, 53, 76, 126, 128–130

Coordination, 3, 31, 32, 84, 129

Cyber security, 3

D

Decentralized, xi, 4, 9–11, 25, 53, 93, 128

Deception, 83, 94, 122, 127

Distributed sensors, ix, 51, 59, 84, 94, 125

Disturbances, ix, 1, 3, 5, 82, 84, 93

Dual-layer, x, xi, 16, 20, 86, 93, 94, 122, 127, 133

Dual-layer task allocation system, 3, 19, 20

E

Efficiency, xi, 3, 31, 50, 125

Evolutionary computation, 28

Execution delay, 21, 76, 126, 131

F

Facility monitoring, xii, 3

Fault tolerant, x, xi, 4, 6, 105, 107, 127, 129

Fire monitoring, xii, 3, 24

Fisher market clearing, 33, 39, 133

Forward-Looking Infrared Radar (FLIR), 95

Framework, xi, 3, 4, 6, 16, 19, 51, 81, 82, 122, 125, 128, 129

G

Genetic algorithm, 33, 42

Greedy algorithm, 38, 53, 55, 68, 121

H

Heterogeneous Distributed Bees Algorithm (HDBA), x, 4, 34, 84

© Springer Nature Switzerland AG 2020
I. Tkach and Y. Edan, *Distributed Heterogeneous Multi Sensor Task Allocation Systems*,
Automation, Collaboration, & E-Services 7,
https://doi.org/10.1007/978-3-030-34735-2

Printed in the United States
By Bookmasters